A STUDY OF WOMEN'S ROLE IN IRRIGATED AGRICULTURE IN THE LOWER VAKSH RIVER BASIN, TAJIKISTAN

DECEMBER 2020

Note:
In this publication, "$" refers to United States dollars.

Cover design by Jasper Lauzon.

Project photos from Alan Clark and Nozilakhon Mukhamedova.

Contents

Tables, Figures, and Boxes

Acknowledgments

This study was carried out under the Asian Development Bank (ADB) technical assistance (TA) project Preparing the Irrigation and Drainage Modernization in the Vaksh River Basin Project.

Yasmin Siddiqi, director, Environment, Natural Resources and Agriculture Division (CWER) of ADB's Central and West Asia Department (CWRD), initiated the study and provided overall guidance and supervision. The report was researched and authored by Nozilakhon Mukhamedova, poverty and gender coordinator (TA consultant) with inputs from Mary Alice Rosero and Maria Lisa Alano (consultants) of the Portfolio, Results, Safeguards and Gender Unit, CWRD. Jelle Beekma of the Water Sector Group of ADB's Sustainable Development and Climate Change Department (SDCC); Prabhjot Khan of the Gender Equity Thematic Group, SDCC; and Gulnora Kholova (consultant), CWRD, provided technical reviews and comments. CWER's Akihiro Shimasaki, Noriko Sato, and Kristine Joy S. Villagracia led the publication process. Zarema Akhmadieva and Husnoro Saidova helped the author during the preparation of the report.

The ADB Tajikistan Resident Mission—headed by Pradeep Srivastava, country director—and the Agency for Land Reclamation and Irrigation in Dushanbe and in Yovon and Jayhun districts extended support in the administration of the study. The water users associations and study respondents provided sound inputs and utmost hospitality.

Abbreviations

ADB	Asian Development Bank
ALRI	Agency for Land Reclamation and Irrigation
ha	hectare
GDP	gross domestic product
I&D	irrigation and drainage
NGO	nongovernment organization
O&M	operation and maintenance
USAID	United States Agency for International Development
WUA	water users association

Currency Equivalents
As of 20 November 2020

Currency Unit: Somoni (TJS)
TJS1.00 = $0.0885
$1.00 = TJS11.3

Executive Summary

The Irrigation and Drainage Modernization in the Lower Vaksh River Basin is a grant-financed project of the Asian Development Bank (ADB). It plans to modernize the Yovon irrigation and drainage (I&D) systems (in this report referred to as Right Tributary) and (priority elements of) the Kumsangir I&D system (Kumsangir) in the Lower Vaksh River Basin, Tajikistan. It aims to increase the capacities of the Agency for Land Reclamation and Irrigation (ALRI) and water users associations (WUAs) for improved water management. The project has a special focus on women's role in irrigated agriculture and will target women land operators.

The ADB regional technical assistance (TA) project Strengthening Gender-Inclusive Growth in Central and West Asia—the Subproject on Feminization of Agriculture in Tajikistan is a separate detailed study that has been undertaken to better understand women's role in irrigated agriculture and water management in the two I&D systems.

The primary objective of the study was to obtain appropriate data for the establishment of baseline indicators for the project. This is to provide the basis for developing socioeconomic and gender-related elements of the project design and impact evaluation. The study provides detailed descriptive information and analysis for subsequent gender-targeted interventions. These will be implemented under the proposed ADB grant-financed project.

The study seeks to deepen the understanding about rural women's roles and existing gender gaps related to land operation and farming; labor arrangements in agriculture and water management; and the feminization process.

The mixed quantitative and qualitative methods study is based on a household survey of 323 respondents and semi-structured interviews and focus group discussions of around 100 respondents. These were conducted from October 2019 to January 2020.

The study showed that the role of Tajik women in agriculture falls under three categories: (i) agricultural wage or daily workers, (ii) women managing kitchen gardens, and (iii) *dehkan* (family-based farm) farmers. These roles may overlap, especially the role of women as caretakers, small agricultural producers within their homes, and visible role as wage or daily workers that are often not recognized in official statistical reports. In 2018, 69% of women in Tajikistan were officially employed in the agriculture sector, compared to 41% of men. Yet, based on subjective perceptions of study respondents as wage or daily workers, women make up over 80% of agricultural labor.

Around 50% of households have at least one person who had left to work outside Tajikistan. Khatlon province has a higher rate of male outmigration (39%) than the national average (36%). These high levels of male outmigration have led to a substantial increase in women's responsibilities in agriculture in addition to managing household tasks.

Female members of the households contribute significant time and effort to family care and domestic work. They perform unpaid jobs such as maintaining hygiene, cooking, and washing clothes. Women are responsible for water allocation and management within the household and its use for household chores and to maintain hygiene. Kitchen gardening is in the women's domain and is viewed as an extension of their role in the house. Based on qualitative interviews, women decide what to plant in their kitchen gardens, independent of their status of heads of the households.

Although women identify remittances sent by men as significant, 42% of men view proceeds from sale of their own produce as the main income driver. Share of income resulting from agricultural production is no more than 45% of total income for almost 75% of respondents. Around 20% of respondents in the Right Tributary I&D have additional nonfarm income. In Kumsangir, only 9% of men and 5% of women have additional nonfarm income. Despite their overburdened role in agriculture, rural women lack access to land, finances, knowledge, and production inputs.

Females comprise only 8.3% of total *dehkan* members of WUAs in the study area. Khatlon province gender-disaggregated data show that of a total cropping area of 281,424 hectares used by *dehkan* farms, only 8% was used by female-led *dehkan*. Distribution of productivity indicators showed that men perform better in growing wheat, maize and other cereals, potatoes, onions, melons, orchards, and beans. Female farmers in Kumsangir perform slightly better as cotton growers than males, and as vegetable growers in the Right Tributary.

About half the women and men said the main factor facilitating productivity was supply of irrigation water. Availability of good quality seeds was also important. Particularly for female respondents, access to information, credit, and machinery services were important factors. For men, these factors were less significant or absent. About 20% of respondents cited climate change as a serious issue.

Women as decision makers were in the minority or absent among *dehkan* farm managers, and therefore were less visible members of WUAs. Water institutions do not have gender-disaggregated statistics about their members and users.

Moreover, women in agriculture are not equal in terms of economic return and employment. Women have less access to knowledge, skills, and information than men and are paid less than men. Women need vocational training and guidance in their work places.

Nearly 80% of male respondents said that there are no problems in production. However, over 50% of female respondents stated problems, including lack of water, bad condition of canals, users at the head of the canal taking a major portion of water, natural drought, and unsatisfactory work by WUAs. Around 20% of female respondents experienced verbal or physical abuse when asking for water.

The recommendations from this study target four key areas of intervention: (i) land tenure security; (ii) access to inputs, knowledge, and finance; (iii) gender mainstreaming in water governing and managing institutions; and (iv) labor participation in agricultural positions. Issues raised and recommendations provided in the report fall into the "(i) women's economic empowerment increased" and "(iv) women's time poverty and drudgery reduced" strategic priorities set in the ADB Strategy 2030 Operational Plan for Priority 2, Accelerating Progress in Gender Equality.

These recommendations are:

(i)   Help women complete the land registration process through practical handholding.
(ii)   Establish agriculture consultation services through mobile phones based in a call center.
(iii)   Improve access to quality inputs, technologies, and machinery.
(iv)   Improve access to financial products.
(v)   Facilitate access to knowledge and information through targeted training.
(vi)   Support capacity building of ALRI for designing a gender strategy with an activity plan for the ALRI gender group, including gender-sensitive staffing.
(vii)   Support ALRI in creating a common database of information for analytical work.
(viii) Ensure water infrastructure and access to irrigation water for households.
(ix)   Ensure availability of separate basic facilities for women hired by farms and farmers.
(x)   Promote alternative agricultural activities for *mardikor* (hired worker) groups.
(xi)   Establish daycare and kindergarten facilities for children of *mardikor* groups.

Chapter I

INTRODUCTION

A. Background

1. Economic Status

Agriculture is Tajikistan's dominant economic activity. Two-thirds of labor resources and half the population of Tajikistan are in the agriculture sector, which accounts for 20% of the country's gross domestic product (GDP) and about 30% of exports.[1] In 2018, earnings from exports of fruits and vegetables totaled $15 million to $20 million.[2]

Tajikistan has a narrow economic base, which is reliant on limited raw products (cotton and aluminum) and remittances from migrant labor. About 95% of Tajik migrant laborers are men from rural areas.[3] Their remittances comprise about 30% of GDP, and the poorest rural households finance about 80% of their annual consumption through these funds. Rapid population growth remains characteristic of Tajikistan, with a steady average growth rate of 2.5%. The population in Khatlon province grew from 1.88 million in 1991 to 3.28 million in 2018, and 70% of them reside in rural areas.[4]

2. Agriculture and Natural Resources

Although freshwater resources are abundant (its annual water availability is 17,000 cubic meters/person/year), Tajikistan has the lowest ratio of irrigated land to population in Central Asia.[5] Only 6% of the country's surface is arable, and part of the arable land is degraded due to salinization and land erosion. About 70% of agriculture depends on irrigation. Most irrigation systems were developed prior to Tajikistan's independence in 1991 and they have now deteriorated or are entirely defunct. Their operation and maintenance (O&M) are inadequate due to a lack of financial resources. Deficiencies in on-farm water management include absence of precision land leveling, suitable furrow lengths for conveying irrigation water, etc. Deteriorated irrigation and drainage (I&D) facilities and poor on-farm water management have led to low crop yields, excessive water losses, water logging, and soil salinization. Cotton and wheat are the predominant crops, with 14% of the cultivated area under vegetables, potato, and cucurbits production.[6]

Limited farm inputs, limited access to extension services, and poor rural–urban connectivity exacerbate low agriculture productivity, making Tajikistan the most food insecure country in Central Asia.

---

[1] World Bank. Agriculture, Forestry, and Fishing, Value Added (% of GDP) (accessed 7 July 2019).
[2] United States Agency for International Development (USAID). Feed the Future Project (accessed 26 August 2020).
[3] N. Mukhamedova and K. Wegerich. 2018. The Feminization of Agriculture in Post-Soviet [Union] Tajikistan. *Journal of Rural Studies*. 57. January. pp. 128–139.
[4] World Bank. World Bank Open Data Database. https://data.worldbank.org/ (accessed 15 July 2020) database 2020.
[5] Agency for Land Reclamation and Irrigation (ALRI). 2019. *Annual Report 2019*.
[6] Government of Tajikistan, Statistical Agency under the President. 2018. *Statistical Collection on Agriculture in Republic of Tajikistan*. Dushanbe.

3.   Land and Water Rights

Since the country's independence in 1991, the key agricultural resources of land and water have undergone structural changes. Land reform efforts resulted in a greater number of individual or commercial farms and a simultaneous decrease in farm size. Land ownership in the country is controlled by the government. Land cannot be privately owned and the government has the authority to expropriate and assign lands based on national goals. However, land legislation allows land use rights. Individuals and legal entities can become primary users of agricultural land and be given usufruct rights through different tenure arrangements based on either indefinite, terminable, or lifetime inheritable uses.

Four main agricultural land tenure types exist: (i) cooperatives or enterprises,[7] (ii) individual/ commercial farms (*dehkan* farms),[8] (iii) presidential plots, and (iv) household plots (kitchen gardens). Rural households typically have access to kitchen gardens. Figure 1 depicts the change in farm types and their dynamics over the past 20 years. The national statistics only provide information on three major types: (i) agricultural enterprises, (ii) *dehkan* farms, and (iii) household plots. A fourth category, *kolkhoze* and *sovkhoze* (collective and state farms), has dissolved, while *dehkan* farms have increased in number.

Figure 1: Dynamics of Farming Units, 1997–2017
(number of farms)

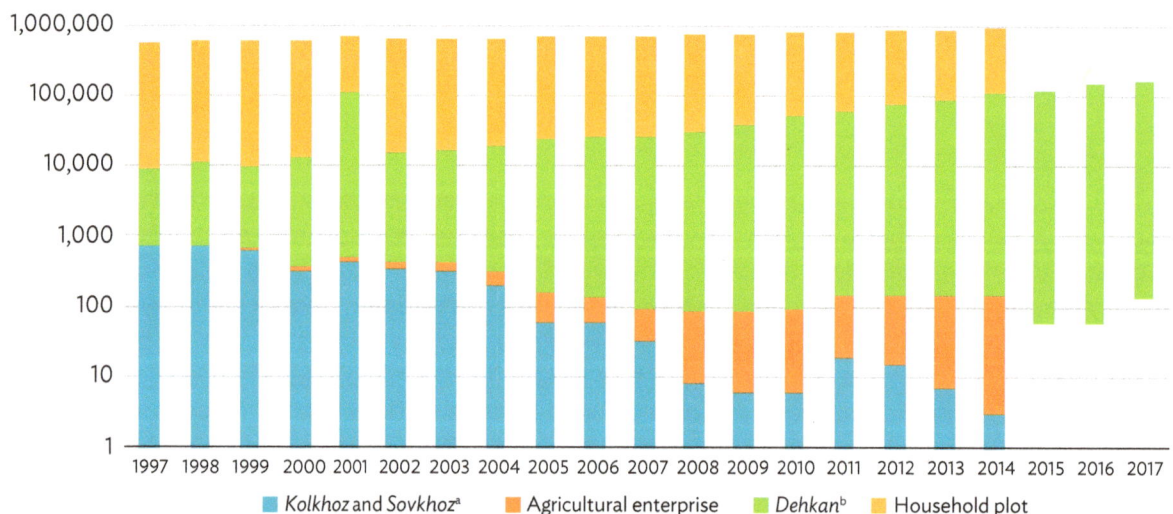

_a Collective and state farm
^b Family-based small-scale enterprise that produces and markets agricultural products using labor of family members on a land plot transferred to the head of family for lifelong inheritable ownership, registered or not registered officially as a legal entity (*dehkan* literally means "farmer")

Note: The data for collective and state farms changed in the statistics provided in the *TajStat Report 2017*, and increased in numbers. Information for households and agricultural enterprises for 2015–2017 are missing in the statistical sources.

Source: Analysis based on data from Government of Tajikistan, Statistical Agency under the President (TajStat). *Gender Indices in Dehkan Farms for 2012–2017*. https://www.stat.tj/en/publications (accessed 16 November 2020).

---

[7]   During the years of the former Soviet Union, cooperatives or farming enterprises used to be collective and state farms, which have shrunk with time.
[8]   *Dehkan* farms are commercial, small- or medium-scale enterprises that produce and market agricultural products using labor of family members or hired workers, registered or not registered officially as a legal entity. *Dehkan* literally means farmer.

Women are a minority among farmers who operate and own agricultural value chains. Of an estimated 57,600 *dehkan* farms in the province, 10,000 (around 17%) were registered as female-headed.[9]

Land. Cooperatives and individual or commercial farms were created on former state or collective farmland or from other land known as the state fund.[10] Agriculture cooperatives or enterprises served as the initial transformation of collective and state farms before individual ownership, and some maintained the same farm structure by keeping the land share owners as members of the cooperatives or staff of the enterprise. Such farms are usually larger than 30 hectares (ha) and may combine plots of several family members, relatives, or groups of farmland share owners who have not yet become legal farmers.[11]

Individual or commercial farms resulted from division and redistribution of collective and state farmlands into shares (*sahm*). People with shares could obtain certificates that provided long-term lease and rights for individual land use. Permanent heritable land use rights are provided to individuals (men or women) and legal entities who can use the rights for a certain term or lifetime (inheritable) basis; and secondary users—individuals and legal entities—who use the land under rent contracts up to 20 years.[12] This allows inheritance and subletting land, but not its sale. Thus, land is not owned by farmers or households[13] but is guaranteed via long-term lease agreements from the government. Many rural people who previously worked in Soviet Union farms received a share and a certificate with the right to use their land plot, while others could not be registered as legal *dehkan* farmers due to the absence of land use rights. Apart from land shares for *dehkan* farms, some additional land was allocated in the form of "presidential lands." Such small plots were distributed by presidential decrees (in 1995 and 1997) to those who had smaller kitchen gardens than the national minimum. Kitchen gardens are owned by most rural households and considered important for food self-sufficiency and a source of income. By the end of 2017, crops grown by households in the Khatlon province exceeded 45% of provincial production. Depending on size, land quality, and water availability, these kitchen gardens may be converted to commercial farming, subsistence agriculture, or not used for cropping activities.

Water. Since the collapse of the former Soviet Union, the water sector has transformed and combines national and basin governance. The Ministry of Energy and Water Resources is the central authority that formulates and fulfills the public policy and regulatory functions for fuel, energy, and water resources. Under the Ministry, the Service for State Supervision over Hydraulic Structure Safety is an authorized executive body that ensures the control of hydraulic structures after their commissioning. The Agency for Land Reclamation and Irrigation (ALRI) manages national, provincial, and district water resources. Water users associations (WUAs) are another actor in water management. WUAs were created with the purpose of transferring irrigation management from state-led farms to noncommercial organizations. Entities entitled to usufruct rights and that have access to agricultural

9   Government of Tajikistan, Statistical Agency under the President (TajStat). 2018. *Gender Indices in Dehkan Farms for 2012–2017.* (accessed 16 November 2020).

10 S. Robinson et al. 2008. Land Reforms in Tajikistan: Consequences for Tenure Security, Agricultural Productivity and Land Management Practices. In R. Behnke, ed. *The Socio-Economic Causes and Consequences of Desertification in Central Asia.* pp. 171–203. Springer Science + Business Media B.V.

11 Cooperative or enterprise type of farms were not the focus of the study and therefore not included in the analysis.

12 Government of Tajikistan. 2002. *Law of Republic of Tajikistan On Dehkan (Farms); and Government of Tajikistan. 2013. Law of Republic of Tajikistan On Cooperatives* (accessed 16 November 2020).

13 "The land, its subsoil, water, air, flora and fauna and other natural resources are the exclusive property of the state, and the state guarantees their effective use for the benefit of the people." Government of Tajikistan. 1994. Constitution of the Republic of Tajikistan. Amended in 2003. Article 13 [2].

lands also become recognized and accounted as farmers, or primary water users, and can become members of a WUA. However, there are cases when WUAs are bypassed and water service agreements are made directly with ALRI.

4.   Khatlon Province

The Khatlon province is located in the south of Tajikistan and is composed of 12 districts with a population of over 3 million in 2019, 53% of whom are women (footnote 9). The majority (64%) of the population of the province is occupied in agriculture of whom 43% are women.[14]

Cotton and cereals are the major crops, occupying 73% of total cropland in Khatlon. The province is also the largest horticulture grower in Tajikistan (Figure 2), producing 42% of the country's total production of fruits and 55% of vegetables (footnote 14).

Figure 2: Vegetable and Fruit Yields by Farm Types, 2005–2016

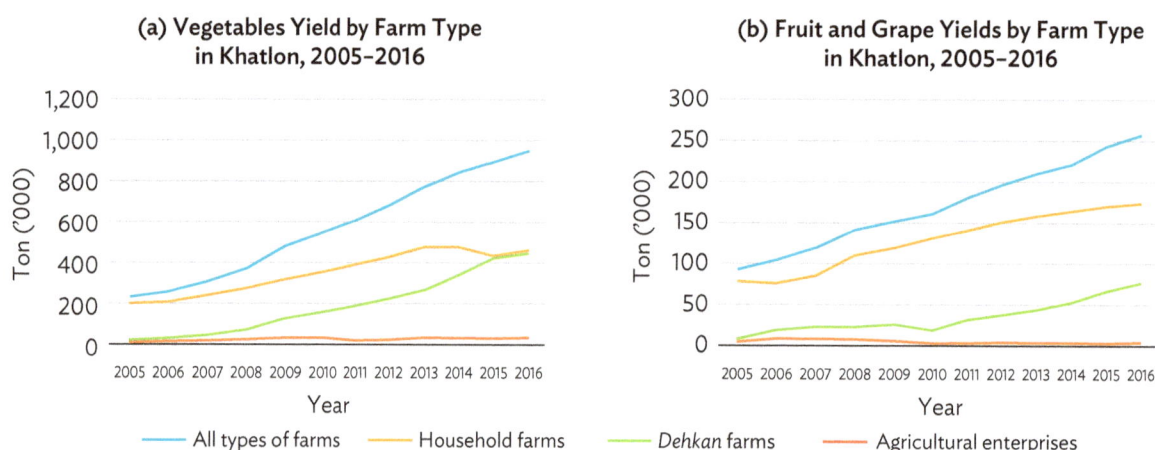

Source: Compiled by the Z-analytics group based on Government of Tajikistan. Statistical Agency under President of the Republic of Tajikistan (TajStat). 2019. Agriculture in the Republic of Tajikistan. Source: Compiled by the Z-analytics group based on Government of Tajikistan. Statistical Agency under President of the Republic of Tajikistan (TajStat). 2019. Agriculture in the Republic of Tajikistan.

The share of households and *dehkan* farms in horticultural production has been significantly increasing over time due to the allocation of more land for horticulture and increasing yields, particularly in *dehkan* farms. However, these crops account for only 20% of the total cropland in Khatlon.

---

[14]   Government of Tajikistan, Statistical Agency under the President (TajStat). 2018. *Labour Market in the Republic of Tajikistan 2017* (accessed 16 November 2020).

B. Study Methodology

1. Purpose of Study

The Asian Development Bank (ADB) will finance a $25 million grant project in 2020 for Irrigation and Drainage Modernization in the Lower Vaksh River Basin. The project will modernize Yovon and (priority elements of) Kumsangir I&D systems in the basin. It will increase capacities of the ALRI and WUAs for improved water management. Women land operators will be targeted in all activities.

This study aimed to understand women's role in irrigated agriculture in the two I&D systems.[15] Specifically, it sought to improve insights about factors affecting agricultural yields and the unique role women play. Results of the study were to inform subsequent gender-targeted interventions under the project.

2. Specific Objectives

The study developed qualitative information using a household survey, in-depth interviews, and focus group discussions. The study identified the following key topics:

(i) socioeconomic baseline survey to ascertain women's status in the household, contribution to irrigated agriculture, and access to land and other resources;
(ii) role in irrigated agriculture: activity location (home garden, main plot for key cash crops, etc.), specific activities and employment type, and land tenure status;
(iii) role in water management: women's role in WUAs and their contribution to decision making on water management aspects, and women *mirob*[16] engagement in operation and maintenance (O&M) of I&D systems;
(iv) aspects of women's agricultural position: work sites, organization, wage issues, modality of payments (cash or in kind), and contractual arrangements (formal and informal);
(v) women's access to knowledge, skills, and inputs for improved production; and
(vi) women's ideas and opinions on what they consider important for improved agricultural productivity.

The study had three stages: (i) an exploratory field visit, (ii) two extensive visits for qualitative interviews and focus group discussions, and (iii) a household survey. Secondary information was collected from yearbooks of the Statistical Agency under the President of the Republic of Tajikistan, WUA monitoring reports, and policy and legal documents. Unfortunately, statistical data requested (officially through ALRI) from district and village authorities were not provided.

---

[15] The study was financed under regional technical assistance Strengthening Gender-Inclusive Growth in Central and West Asia under the Subproject on Feminization of Agriculture in Tajikistan.
[16] Female water masters involved in irrigation services or irrigation processes within farms.

3.   Stages of Survey Design and Implementation

Stage I

Since the study was closely connected with planned rehabilitation and modernization activities in specific I&D systems, a hydraulic approach was taken to select the study site. To focus on the Lower Vaksh River Basin and selected schemes, the study surveyed beneficiaries who received water only from Vaksh River Basin service areas. Further details are in Appendix 1.

Stage II

Data collected were from two primary sources: a quantitative study (household survey) in irrigation systems and districts identified by ADB and ALRI, and qualitative instruments—focus group discussions and individual, key informant interviews. Women respondents were sought specifically as part of the sampling methodology. Also, information (interviews of experts and secondary data) was collected from communities, ALRI, WUAs, international organizations, and nongovernment organizations (NGOs) working in I&D service areas.

Stage III

The main goal of the baseline household survey was to collect primary data on the extent of feminization of agriculture in the selected I&D systems. A questionnaire was designed based on the ADB pre-approved outline and were divided into the following 10 sections:

(i)   general information;
(ii)   household structure, demographic characteristics, and farm ownership;
(iii)   land tenure;
(iv)   agricultural data;
(v)   irrigation data;
(vi)   decision-making in the household;
(vii)   institutional aspects;
(viii) labor aspects;
(ix)   access to information and farming skills; and
(x)   household income and expenditure.

Chapter II

AGRICULTURE AND IRRIGATION SITUATION AND CHALLENGES

A. Broad Situational Analysis, National Level

1. Agency for Land Reclamation and Irrigation

1.1. General Background

ALRI is the central executive body in Tajikistan regarding land reclamation and irrigation, carrying out the functions of developing a unified state policy and regulations in the field of land reclamation and irrigation systems, use and conservation of water facilities, provision of irrigation water, and water conservation.[17] The mandate of ALRI includes specific roles related to the O&M of water facilities; designing and building new hydraulic structures, preventing flood risks, monitoring land reclamation conditions, and water management and water resources use for irrigation. Another important role of ALRI and its branches is to support and monitor WUAs.

ALRI has five regional offices and 14 branches responsible for various activities and locations, based on the former Soviet Union administrative district organizational structure. Since the independence of Tajikistan, they have been under institutional changes due to new land and water policies, introduction of new farming systems, and establishment of the WUAs.

ALRI produces hydraulic data and reports on water source intake and water use, irrigated area (gravity and machine), water and electricity use, and total prices at district levels. Sample ALRI reports are in Appendix 2.

1.2. Staffing Profile and Women's Representation

At the end of 2019, women accounted for 12% (758 persons) of ALRI staff.[18] Women are employed mainly as technical staff: accountants, statisticians, secretaries, cleaning staff (68%, 512 persons), and specialists such as on WUA monitoring, pump station monitoring, shore protection, and mudslide protection (22%, 140 persons); 3% are chief specialists, and 3% are division heads. Water experts claim that there are not so many women in the agency. Women are not always accepted in managerial positions where they are equally eligible as specialists or managers.[19] Usually, such positions are offered to and occupied by men, who in turn enjoy more social power and opportunities to participate in projects. No initiatives or measures are taken to promote and support female hydro technicians and ameliorators.

---

17   ALRI (accessed 1 January 2020).
18   As of December 2019, the total number of ALRI staff is 6,132.
19   Personal interview with water experts. October–November 2019. Dushanbe.

Based on several government initiatives and legislative acts accepted by the Government of Tajikistan, ALRI established a gender group of three persons in 2015: Deputy Director of the Agency (male), and two female staff members based in the head office. The gender group plan of activities includes establishing similar groups in most of its sectoral branches, organizing meetings, and participating in training, seminars, roundtables, and career development courses on gender issues related to the water sector. District branches in the study areas did not have gender groups. According to ALRI, the establishment of a gender group was not financed by any development partner. No previous international project related to water governance or irrigation supported its activities or contributed to gender capacities of ALRI. ALRI is accountable to the Committee of Women and Family Affairs under the Government of the Republic of Tajikistan. However, ALRI is required to submit only the annual number of female staff in the whole ALRI system to the Committee for Women and Family Affairs.

The Committee of Women and Family Affairs, being the central executive body, implements a state policy to protect the rights and interests of women and the family, creating equal conditions for the realization of their rights, achieving gender equality, expanding participation of women in solving socioeconomic problems, and management of state and public affairs.[20] It is expected that the National Gender Strategy will be prolonged beyond 2020 or a new 5-year plan of activities developed, which would include indicators to monitor gender equality nationwide, including ALRI. According to the National Strategy, ALRI should address such issues as female understaffing in management positions by establishing quotas, sensitizing staff on gender issues, and promoting gender-sensitive staffing, as well as educating more women hydrotechnical specialists, etc.

According to several legislative resolutions and decrees,[21] there should not be any legal barriers for women to access leading or higher positions; public sector agencies should have one deputy female director, but this is not followed in ALRI. "The presidential office that monitors the water sector indicated that there are no women specialized in irrigation and land reclamation of the required experience level, but it is expected that there is a search for such a candidate."[22]

Women in ALRI branches are involved in senior positions marginally or not at all. The interviewed female experts expressed their need for capacity building and insuring roles for female specialists and managers in the ALRI system. Other ministries have reconsidered their human resource management toward gender-sensitive hiring processes and appointing women management positions. The key respondents noted that, unfortunately, the organizations within water governance avoid female managers, heads or deputy heads, as the male managers deem women incapable of holding such positions. Especially in rural areas, as primary multipurpose users of water, women are often ideally suited to drive change in the design and maintenance of water systems, water distribution, and policymaking.

---

20   Committee for Women and Family under the Government of the Republic of Tajikistan (accessed 10 January 2020).
21   Government of Tajikistan. 2006. Resolution No. 496 (1 November 2006) Government Programme *On Co-Mentoring, Selection and Job Placement of Leading Personnel of the Republic of Tajikistan From Among Gifted Women and Girls for Years 2007–2016*; Government of Tajikistan. 2001. Resolution of the Government of Republic of Tajikistan No. 391 (6 August 2001) Government Programme *On Main Directions of Government Policy on Providing Equal Rights and Opportunities for Man and Women in Republic of Tajikistan for 2001–2010*; and Government of Tajikistan. 1999. Decree of the President of the Republic of Tajikistan *About Measures on Increasing of Status of Women in Society*.
22   Personal interview with water experts. October–November 2019. Dushanbe.

2. Water Users Associations

During the Soviet period, water governance at the farm level was centrally organized and managed by local water management authorities. Primary control lay with the Ministry of Irrigation and Water Management whose policies were executed through provincial administrative branches. Provincial officials, in turn, worked with district irrigation offices to operate and maintain primary and secondary canals.[23] After independence and the initiation of reforms, farm-level management of water resources was transferred to WUAs.

2.1. Responsibilities

According to the current legislation, WUAs are nonprofit and self-governed organizations of water users based on membership and are created voluntarily by legal authorization. They are responsible for the operation and maintenance of inter-farm level canals within hydrological boundaries,[24] and have the right to operate and maintain irrigation and drainage networks at the inter-farm and in some cases at the intra-farm level.

Most WUAs were established through programs led by the World Bank in 1999, and later Swiss Agency for Development and Cooperation, Deutsche Gesellschaft für Internationale Zusammenarbeit (GIZ), ADB, Helvetas International, and the United States Agency for International Development (USAID), creating more than 400 WUAs in less than 2 decades. WUAs were expected to (i) fill an administrative gap, (ii) improve cost efficiency, and (iii) facilitate services that are more responsive by leveraging local knowledge and participation.

According to ALRI statistics (2019), 392 WUAs exist in Tajikistan, with 133 WUAs in Khatlon province. Of the 21 districts of Khatlon province, 9 districts use the Vaksh River as a water source (Appendix 3).

Each WUA produces monthly monitoring reports that consist of institutional data, WUA profile, water use, financial data, inventory of equipment and infrastructure, and training and agricultural characteristics of WUA members.

Table 1 shows the number of WUAs within the study canals, Right Tributary and Kumsangir, from four district monitoring reports.

---

23   J. Sehring. 2006. *The Politics of Irrigation Reform in Tajikistan*. Discussion Papers/Zentrum für internationale Entwicklungs-und Umweltforschung.

24   United Nations Economic Commission for Europe. 2012. *Roadmap of the National Policy Dialogue on IWRM in the Republic Tajikistan*. Dushanbe, Tajikistan; J. Sehring. 2009. *The Politics of Water Institutional Reform in Neopatrimonial States: A Comparative Analysis of the Kyrgyz Republic and Tajikistan*. 1st ed. Wiesbaden: VS, Verlag für Sozialwissenschaften (Politik in Afrika, Asien und Lateinamerika).

Table 1: Inventory of General and Project Focus Water Users Associations in Relevant Districts

District	Number of WUAs in Each District	Number of WUAs in the Vaksh River System
Yovon	9	4
A. Jomi	9	2
Khuroson	4	4
Jayhun (Kumsangir)	8	7
Total:	**30**	**17**

WUAs = water users associations.

Source: Agency for Land Reclamation and Irrigation (ALRI). 2019. *Report of Inventory of General and Project Focus Water Users Associations.* Compiled by the author.

The directors of all but 1 of the 30 WUAs are males. The female WUA director has been working in her position for 3 years and mostly deals with payments from farmers from the village authorities responsible for collecting fees from the households. However, most neighboring WUA directors in the district do not consider her to be competent in her position and believe that she was assigned a nominal position.

2.2.   Water Management Responsibilities

i.   Membership Structure and Roles

WUAs in the project area have similar staff composition according to their charter: director, manager, hydro technician, accountant, secretary, and *mirob*. However, such staffing is not consistent. In some cases, the responsibilities overlap; in others they might be absent. Moreover, WUAs can open new staff positions. Female staff are usually hired in WUAs for more conventional secretary or accountant roles, but may also be invited to perform new roles such as that of *mirob* or irrigation fee collector.[25]

Mirob workers exist on various levels: as water supply engineers in ALRI branches, as water masters (not necessarily water professionals) in WUAs, and those who are not particularly recognized and are excluded from the formal "water institutional hierarchy." According to ALRI staff, the prescribed norm for a *mirob* is to serve 5 kilometers (km) of an irrigation canal. Currently, *mirob* and staff of ALRI branches and WUAs serve over 10 km of canals. *Mirob* workers receive only around TJS400 or $40 a month. The WUA director receives TJS800 every month.

ii.   Payment Systems

Payments by water users in their contracts with WUAs are of two types. The first is an annual membership fee set by each WUA of about TJS40/ha annually on average for Tajikistan. In Right Tributary canal WUAs the fee is TJS27 to TJS55, and in Kumsangir, it is TJS50 to TJS65 per hectare per year. The membership fee for households including water service fees is TJS2 to TJS5 for 0.01 ha/year.

---

25   Focus group discussion with Yovon WUA chairman, Right Tributary I&D system. October 2019.

The second type of payment is the irrigation service fee. Such fees can be based on crop type and land size or, if there is a measuring device such as a water meter by cubic meter (m^3) used. The meter fee is set by the state and in 2018 was TJS0.02/m^3 for both farmers and households.[26]

2.3.   Demographics and Characteristics of Water Users Association Members

More than 203,000 people live in the two study schemes, consisting of various land users. Appendix 4 summarizes WUA information for the two I&D systems. Over 40% of the irrigated area is occupied by *dehkan* farms. The number of farmers in the Right Tributary system with more than 30 ha of land is twice that in Kumsangir I&D system. According to WUA bylaws, only legal farming entities can become members of WUAs and their irrigation needs are prioritized. Households that possess only kitchen gardens cannot become members and thus, have no formal voice in water management decision-making processes. WUA directors state that local governments (*jamoat*) represent the interests of kitchen garden households and are often responsible for irrigation service fee collection from the households. However, the power and weight of representation of households by local governments are weak and insignificant compared to those of WUA member farmers.

Female household heads in the study canals comprised only 6.6%–6.9% of the total. However, detailed analysis on WUAs for the same indicator shows a wider variation (0.9%–14.4%). The higher figure probably includes households with outmigrant male absentee members, which is estimated at 68% of total households, and where women often become the decision makers.[27]

The share of male and female WUA members varies considerably, being generally below 10% in Kumsangir canal, while over 20% in a few Right Tributary WUAs. Relatively higher representation of women in these WUAs may be connected to land access and migration patterns or presence and support from international or local NGOs.

Around 87% of all female WUA members in the Right Tributary canal system and over 94% of female members in Kumsangir canal system are represented by *dehkan* members (Table 2). However, only 8.3% of total *dehkan* members of WUAs in the study area are females.

Table 2: Sex-Disaggregated Number of *Dehkan* Farm Members of Water Users Associations in Project Sites

Project Site	Number of *Dehkan* Farms	Number of *Dehkan* Farms Led by Females	Portion of *Dehkan* Farms Led by Females (%)	Portion of *Dehkan* Farms Led by Males (%)
Right Tributary I&D	2,611	298	11.4	88.6
Kumsangir I&D	2,483	127	5.1	94.9
Total in all WUAs	5,094	425	8.3	91.7
Average in project area	2,547	213	8.3	91.8

I&D = irrigation and drainage, WUA = water users association.

Source: Author's analysis and compilations based on monitoring reports of WUAs of Yovon, A. Jomi, Khuroson, and Jayhun districts.

[26]   Expert interview with ADB's Pyanj River project manager Anvar Murodov. October 2019. Dushanbe, Tajikistan.
[27]   ADB. 2020. *Women's Time Use in Rural Tajikistan*. Manila. June. http://dx.doi.org/10.22617/TCS200167-2.

2.4.   Cropping Patterns

Agricultural characteristics of WUAs provide interesting insights into crop production patterns within each scheme, although not sex-disaggregated. WUAs within the Right Tributary canal plant mainly cotton and wheat, in agreement with the unwritten order of planting 60% of land with cotton.[28] Agricultural data from Kumsangir canal WUAs, although having a 30% unwritten quota for cotton, show significant diversification into vegetables, cucurbits, and fodder (Table 3), and also fruit trees and vineyards.

Table 3: Top 5 Crops Sown in Right Tributary and Kumsangir Irrigation and Drainage Systems, 2019

Right Tributary Canal		Kumsangir Canal	
Crops	Area (ha)	Crops	Area (ha)
Cotton	7,040	Cotton	2,426
Wheat	4,014	Cucurbits	1,702
Cucurbits	1,201	Vegetables	1,490
Fodder	761	Wheat	1,333
Vegetables	393	Fodder	1,051

ha = hectare, WUA = water users association.

Source: Author's analysis and compilations based on monitoring reports of WUAs of Yovon, A. Jomi, Khuroson, and Jayhun districts.

The productivity in both canal WUAs for vegetables, cucurbits, and potatoes is high, but much higher for Kumsangir canal WUAs (Figure 3). Jayhun district has favorable climatic conditions and rich soil that does not require extra fertilizers. Gardening, especially citrus trees, is intensive, with high yields also beyond the regular agricultural season (late autumn and winter), providing opportunities to sell produce to neighboring countries (mainly Kazakhstan and the Russian Federation).

---

[28]   Informed by over 80% of the respondents during qualitative interviews in the Right Tributary canal, Tajikistan. October–November 2019.

Figure 3: Average Production and Productivity in Water Users Association Member Farms

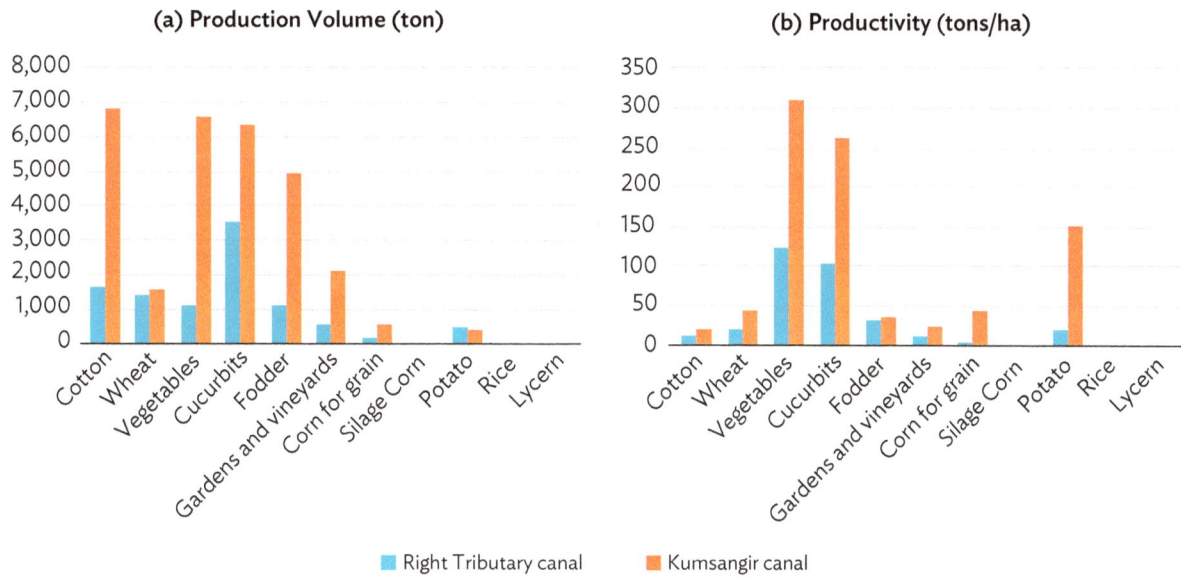

(a) Production Volume (ton)

(b) Productivity (tons/ha)

Right Tributary canal   Kumsangir canal

ha = hectare, WUA = water users association.

Source: Author's analysis and compilations based on monitoring reports of WUAs of Yovon, A. Jomi, Khuroson, and Jayhun districts.

Chapter III
GENDER ANALYSIS OF SURVEY AND INTERVIEW RESULTS

A. Overview of Gender Issues in Rural Tajikistan and Khatlon Province

The official minimum salary in Tajikistan was TJS400 (equivalent to $40) per month in 2019. Low salaries and the scarcity of jobs in rural Tajikistan drive over 800,000 people, mostly men, to migrate out of the country in search of employment. Over 95% of such migration is to the Russian Federation.[29] Khatlon province has a higher rate of male migration (38.9%) than the national average (35.7%).[30] Such high levels of male outmigration have led to a substantial increase in women's responsibilities in agriculture in addition to managing household tasks.

Women's participation in the labor markets of formal and informal jobs is mainly due to multiple demands on their time and energy where no family, community, or state-run support services exist.[31] Informal jobs are those not registered or properly accounted and do not provide an official working status or social protection (pensions and social support for children); they do however, provide some flexibility and additional income, which is at times the main source of household income.[32] This flexible and informal labor supply seems to suit many production businesses that need low-paid workers, such as agriculture.[33]

The outmigration of males in the agriculture sector has led to increased involvement of female labor. The result is, as in other transition countries, feminization of agriculture "an increase in women's participation rates in the agriculture sector, either as self-employed or as agricultural wage workers; in other words, an increase in the percentage of women who are economically active in rural areas."[34] In this study, the definition of feminization is extended to include unregistered informal jobs and unpaid labor.

Tajikistan ranks 123rd of 148 countries on the global gender gap index, scoring lowest on women's economic participation and political empowerment.[35] The jobs women hold have little protection, security, or earnings. Higher female participation is a sign of a wider spectrum of labor opportunities as well as greater sensitivities to economic, social, and political events and the growth of women's power as decision makers.

---

[29] USAID. 2010. *Gender Assessment USAID/Central Asian Republics*. Washington, D.C.

[30] C. Francisco and M. Bakanova. 2014. Tajikistan: Reinvigorating Growth in the Khatlon Oblast. *Europe and Central Asia Knowledge Brief* No. 70. Washington, DC: World Bank.

[31] A. S. Wharton. 2013. *The Sociology of Gender: An Introduction to Theory and Research*. Wiley, Hoboken, N.J.

[32] International Labour Organization. 2010. *Migration and Development in Tajikistan: Outmigration, Return and Diaspora*. Moscow. September.

[33] G. Standing. 1999. Global Feminization through Flexible Labor: A Theme Revisited. *World Development*. 27 (3). pp. 583–602.

[34] E. Katz. 1995. Gender and Trade within the Household: Observations from Rural Guatemala. *World Development*. 23 (2). pp. 327–342.

[35] World Economic Forum. 2018. *The Global Gender Gap Report*.

1.   Women's Roles and Issues in Agricultural Production

Poverty is severe in rural areas since there are limited employment and income opportunities outside the agriculture sector. The burden on women increases as they try to feed their families while engaging in agricultural activities, managing the home, and childcare. Low education and lack of access to extension advice and inputs prevent women from improving crop productivity.[36] Females are occupied in tedious and difficult work, often during very hot or extremely cold weather, but receive two-fold lower wages than men in the agriculture sector. Changes in gender composition of the agricultural workforce justify the need for gender-inclusive and gender-targeted agriculture and water management programs.

Water institutions do not have sex-disaggregated statistics on their members and users other than land ownership. Khatlon province gender-disaggregated data show that of the total cropping area (281,424 ha) used by *dehkan* farms, 8% was used by female-led *dehkan* farms (footnote 36).

Production volumes in the study canals follow similar cropping patterns: 45% of all produce was grain and wheat, followed by vegetables (18%), cotton (13%), and fruit trees (12%) (footnote 36). Productivity data for the main crop categories indicate that female farmers show productivity results equal to or higher than their male colleagues.

The productivity of female-headed farms is almost triple that of males for fruit trees, which require little tending; almost equal in wheat and maize; and less in other grains and cotton production. Women are less productive in high-value crops such as vegetables and cucurbits. This may be connected with access to finance for additional labor and inputs. While these results might be dampened by the quality of seed, fertilizers, water availability, or small farms versus large farms, feminization consists not only of increased female workers in agriculture, but also their capacity to produce higher yields and develop better managerial skills than male farmers.

2.   Labor Participation

In 2018, about 69% of women were officially employed in the agriculture sector, compared to 41% of men (footnote 9). Tajikistan has a relatively unique situation in that women are heavily engaged throughout the entire crop production process. Women's involvement (owners, users, or workers) in agriculture formally counts only when they are registered as legal entities or farm workers. Women are also heavily involved in unpaid family labor—they take care of a multigenerational family and are responsible for the home garden and securing water and food. Although the agriculture output of kitchen gardens significantly contributes to production and food security, it is presented without sex attribution in statistics and national reports.

The feminization of agriculture differs in various regional, village, and family contexts and may have both positive and negative elements. The involvement of women in collecting fees and their direct participation in irrigation activities showed how feminization also led women into support service jobs. Female *dehkan* farmers who handle farming on their own are called "strong" and "warriors" by local

---

[36]   M. C. Buisson. 2018. What Does Male Out-Migration Mean for Women in Tajikistan's Agriculture? *Agrilinks*. 18 June.

populations,[37] as if they are aberrations in a male domain. Women's involvement in previously male-only occupations illustrates how they have become empowered as decision makers and participants in a wide range of management activities. Yet, the feminization process, especially for women in daily agricultural jobs, is accompanied by low wages and an informal job status that excludes them from public social security benefits (footnote 3).

3.   Water Users Associations Access and Participation

Water services are beleaguered by deteriorating infrastructure, weak management, limited staff, high nonrevenue, unaccounted water losses, and poor communication with members and other water users. Half the female respondents in the study recognized that there are issues in water service delivery related mainly to lack of water, bad conditions of canals, and environmental conditions.

Feminization can be observed here too, with women assuming traditionally male-dominated occupations. The long, seasonal male outmigrations led women to take over their roles and such activities as cleaning irrigation canals and collecting water service fees, thus replacing WUA *mirob* workers. Cotton growing, tilling, and harvesting of vegetables and fruits, which in Soviet Union times and Tajikistan's early independence were performed equally by men and women, are now done mainly by women. Although men do not oppose women taking male-dominated jobs, such positions, in the opinion of study respondents, are "not a woman's job," and should be in the male domain.

4.   Household Decision-Making

Conventionally, males conduct all the decision-making within Tajik households. However, due to male outmigration, women have become de facto heads of the household and decision makers. Land plots may be left to be operated by female heads, given for use of close relatives or rented out. In some cases, males formalize the land rights in the name of their wives to prevent additional problems during their absence. Male respondents assume that authorities may treat female farmers better, especially if the actual land rights owner is a seasonal labor outmigrant. This arrangement may be considered beneficial to women, but men or other family members may still keep control and decision-making power. In practice, female *dehkan* farmers make most of the production decisions alone or with the support of male family members, especially on buying quality seeds and fertilizers. At the same time, intra-household hierarchies among female members exist. Elder women are best positioned to make decisions that are accepted by other household members.

---

[37]   Observations and interviews with male and female community members and farmers from Northern and Southern Tajikistan. October–November 2019.

B. Productive Roles and Contribution to Household Economy

1.   Characteristics of Survey Respondents

The socioeconomic baseline survey covered 323 households with 110 female and 213 male respondents. Membership of households averaged 7 persons in both I&D systems. In one household are multigenerational or several families, often consisting of women, children (approximately three per household), and elderly due to the high rate of male outmigration.[38] The average age of respondents was 50, ranging between 25 and 91 years.

Fewer females (65%) than males (88%) indicated finishing at least secondary[39] grades. Male respondents have better access to special or technical and university education. Only 53 respondents, almost equal numbers of males and females, had specialized agriculture background, meaning they either finished a vocational school or university.

The majority of both male (88%) and female respondents (78%) identified themselves as *dehkan* farmers and kitchen gardeners. Around 20% of female respondents, but only 2% of males, were paid farm workers, including hired seasonal workers. Some men (5%) and women (1%) were farm managers. The results were similar for both I&D systems.

Around 50% of households in the sample have at least one person who has left to work outside of Tajikistan (149 migrated members). Qualitative interview results[40] suggest that around 80% of men left the country for wage work. Women also outmigrate, mainly to the Russian Federation, for various job opportunities. Outmigration of male labor forces other family members, especially women, to look for additional income opportunities because remittances are low or not used for daily family needs, or considered to be unstable by the households. For example, because of a recession in the Russian Federation in 2014 to 2015, remittance inflows to Tajikistan dropped from 36% of the GDP in 2014 to 28% in 2015. A similar downturn due to the 2020 coronavirus disease (COVID-19) pandemic is taking place. The economies of the Russian Federation and Kazakhstan have been hit hard by the plunge in oil prices, which triggered the start of a recession in these countries. The numbers of returning migrant laborers rose sharply in February and March 2020, including from Kazakhstan and the Russian Federation, accounting for more than 90% of migrants.[41] The pensions of older members are usually not enough to sustain families with several children, and any decline in remittances challenges a family's food security.[42]

The rural populations in both I&D systems identify remittances and agricultural production as their main income sources. Only 11% of respondents had agricultural businesses other than crop production and these are: machinery rental services, livestock and poultry, greenhouses, and lorry transportation services. There are differences between male and female respondents and their perceptions about the role of the opposite sex. Responses to the question "Who earns what type of income?" elicited

---

[38]   Children under 16 and elderly above the pension age of 65 years.
[39]   The results include secondary grades 9 to 10 or 11, secondary special and technical, higher education (university level). The average age of women in these categories is 48 and for men it is 53.
[40]   Focus group discussions with female *mardikor* groups, Kumsangir I&D system, October 2019.
[41]   *Reuters*. 2020. Tajikistan Says Migrants Returning From [the Russian Federation], Kazakhstan. 17 April. https://www.reuters.com/article/us-health-coronavirus-tajikistan-economy-idUSKBN21Z19H.
[42]   C. Oriol. 2018. *Country Study: Tajikistan*. World Bank and the Eurasian Center for Food Security.

opinions on whether men, women, or both earn certain kinds of income. In the Right Tributary system, 82% of the 67 female respondents perceived that men earn money abroad and send remittances, while 28% perceived that men earn money through "selling own produce." Thus, in the Right Tributary system, female respondents rely heavily on remittances. They also depend more on pensions. Women also consider that their income is formed by selling produce (13%), yet the men perceive this income contribution as much less (1%). Earnings of men in the opinion of both sexes are also complemented by nonagriculture activities and businesses.

In the Kumsangir system, more diversity of income and less migration are evident. Although women still identify remittances sent by men as significant, 42% of men view proceeds from the sale of their own produce as the main income driver. Nonagriculture business is almost nonexistent. The concentration of agriculture in Kumsangir can cause income vulnerability, especially when market dynamics and availability of irrigation water or other production inputs change.

For 35% of respondents in both I&D systems, their family income can satisfy "only food" items while 29% reported that their income can cover food and basic necessities: water, electricity, and transport. Overall, respondents in the Right Tributary have more income for nonfood items than those in Kumsangir.

The share of income from agricultural production was no more than 45% of total income for almost 75% of respondents. Male and female respondents in the Right Tributary had almost equal distribution of agriculture share; males had a greater share than females of agriculture income in Kumsangir.

Besides crop production, agriculture income is gained from cattle (males = 30.5%; females = 45.5%), sheep (males and females = 15%), and goats (males = 11.7%; females = 7.3%). The Right Tributary showed higher variation of livestock income than Kumsangir.

Around 20% of respondents in the Right Tributary had additional nonfarm income. In Kumsangir, only 9% of men and 5% of women had additional nonfarm income.[43]

The survey showed that with respect to the characteristics of respondents by their activities (Figure 4), female respondents worked mainly as caretakers, family farm workers, or were retired. Fewer women work as farm managers or are unemployed. "Family farm employee" is a prevalent answer for men, which means that the family member is working on a family farm and may or may not receive a salary. Among households members are many "retired" (on pension) and "unpaid domestic and care work" categories. Some men work in education, state or public sectors, or as nonfarm business employees.

---

43   Nonfarm income is related to any other type of income received by respondents beyond their farm income. This may include public jobs or private businesses (shops, taxi driving, etc.).

Figure 4: Main Status or Activities of Respondents, Sex-Disaggregated

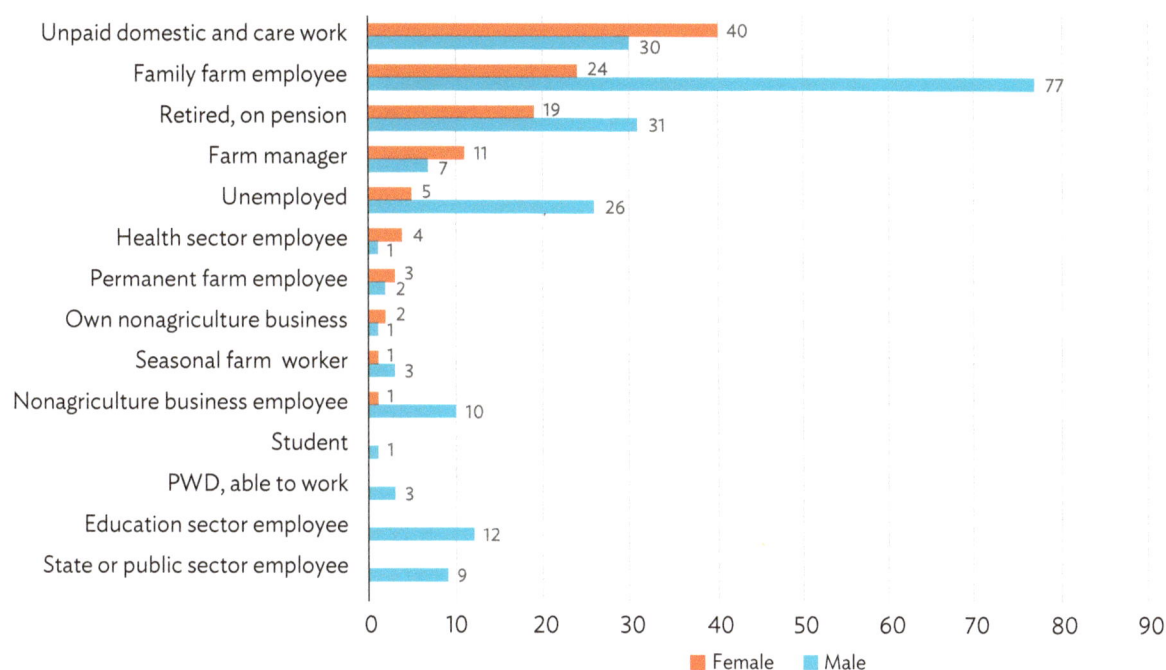

PWD = person with disability.

Note: "Permanent farm employee" may have a share, but is not a decision maker in the farm. "Non-agri-business employee" is different from "skilled labor worker" category as skilled worker includes informal types of work and/or can be specific to construction works.

Source: Data analysis based on individual household interviews (n = 323). November 2019.

C.   Role and Contribution in Irrigated Agriculture and Other Productive Activities

1.   Access to Land

Survey respondents use six types of land for agricultural activities: household kitchen gardens, *dehkan* farm plots, presidential, additional (*yorirason*),[44] and subleased rented and sharecropped lands. Table 4 shows the area of each land type. The irrigated area comprises 94% of the total area of 871 ha. *Dehkan* farms make up 762 ha (87%); those in the Right Tributary are almost double the area of those in Kumsangir. Land used by households is substantially less, only about 20 ha in Kumsangir and 28 ha in the Right Tributary. Rented and presidential land share is around 3% in both tributaries. Only 5 respondents rented land, and 121 respondents answered that they have presidential plots, which average about 0.3 ha in both systems.

---

[44]   Support (*yorirason*) lands are similar to presidential lands. Provided for families who did not have more than 0.15 hectares of kitchen garden plots for agricultural activities. Provision of such land was not restricted to inhabitants of the district or province. Therefore, some people who received *yorirason* lands far from their places of residence turned them into *dacha*, place for recreation or rented them out. The land quality and irrigation services for such lands may be absent or poor. This category of land rarely appear in the national statistics as lands relevant to agricultural production.

Table 4: Survey Respondents' Total Land Area by Type of Land
(ha)

Canals		Total area of land types of respondents, ha			
	Households	Presidential	*Dehkan*	Additional	Subleased
Kumsangir	20	6	261	1	5
Right Tributary	28	9	501	35	6

ha = hectare.
Source: Data analysis based on individual household interviews (n = 323). November 2019

Many respondents rented subleased land according to the current land code. Rent agreements should be registered with a notary and are subject to high taxes in the opinion of respondents of the interviewees. Thus, sublease or sharecropping arrangements are rarely admitted during the surveys, but acknowledged during interviews.

In the Right Tributary, the cost of subleased land was estimated at TJS700/ha to TJS2,000/ha. In Kumsangir where the land is more productive, less risky with regard to both agreements and harvest loss, and in demand, subleases cost TJS5,000/ha to TJS7,000/ha. Women rarely sublease land.

The 6% of nonirrigated land is used for wheat, alfalfa, and other fodder crops in the Right Tributary canal. The average plot size of different types of land is shown in Table 5. The largest land areas are used by *dehkan* farms, of which male respondents have greater areas than female. The size difference of *dehkan* farms between males and females in Kumsangir is 0.35 ha and in the Right Tributary 1.6 ha. The wider gap in land size available for males and females indicates that larger land areas are allocated or leased to male farmers in the Right Tributary. This trend is connected with cotton and wheat planting quotas, which are higher in the Right Tributary I&D, and remaining farmers with several share owners. The tremendous difference of the average size of additional lands allocated to people from the capital and its outskirts may be due to closer access of Yovon district than Jayhun district to the capital, Dushanbe.

Dushanbe is also a big market for agricultural produce and off-farm job opportunities. Qualitative interviews conducted with female wage workers in both I&D systems confirm the survey results. Females have certificates or land use rights for smaller plots (0.05–0.1 ha). Some women received their land shares through the de-collectivization process or inherited from their parents. However, not all were able to get full rights to use the land. Many women still have their certificates with the farm manager who has the full right to decide on production and receive the benefit. The shareholders (*sahmdor*), can legally take out their land from the large collective farms, but there are many uncertainties as to which land plot will be provided and obstacles in registering as entrepreneurs or *dehkan* farmers. Women shareholders are also involved in other types of wage works. Similar to northern Tajikistan, shareholders working in a collective farm may receive a salary, but since such salaries are low, farm managers provide additional in-kind payments that contribute to the shareholders' food security (footnote 3). In-kind payments may include crops produced in the farm or basic food items such as rice, wheat, vegetable oil, fodder, flour, and firewood.

Table 5: Average Plot Size by Types of Land by Gender
(ha)

Land Type	Households		Presidential		*Dehkan*		Additional		Subleased	
	Male	Female	Male	Female	Male	Female	Male	Female	Male	Female
Kumsangir	0.15	0.15	0.04	0.05	2.1	1.75	0	0.02	0.03	0.05
Right Tributary	0.15	0.14	0.05	0.03	3.2	1.60	0	0.50	0.02	0.05

ha = hectare.
Source: Data analysis based on individual household interviews (n = 323). November 2019.

The average size of household plots is almost equal in the two canals for both sexes. The average plots on presidential and rented land range between 0.02 and 0.05 ha for both systems.

2.   Women's Access to Land

The findings of this study show that despite their dominating role in agriculture, rural women lack access to land, finances, knowledge, and production inputs. Women do not have equal access to land in terms of land size. Legally, men and women have equal access to land,[45] but actual possession by women is very low. Women have bureaucratic barriers in validating their land certificates and obtaining access to their land plots. This fact can be tracked not only from official statistics but from numerous interviews and discussions with the rural population. In our sample there are 43 purposefully chosen female *dehkan* farms, of which 40 identified as farm owners and land operators. Yet, interviewees interpreted female farm ownership as nominal and said that in most cases the farms were actually managed by male relatives. This occurs when farms are managed by husbands, sons, or other relatives, or farmlands are rented out. Men and women hold different views about women's access to land. The majority of interviewed male *dehkan* farmers did not have much confidence that women could successfully run farms. Women are not excluded from possessing land in legal terms; however, in practice, they may be restricted in accessing and managing land, water, and other agricultural services.

Women's opinions vary: some follow the same reasoning as men, others are afraid of failure or loss of investments. Some would be eager to run a farm independently if only given access to land, inputs, and knowledge.[46] During the in-depth interviews, women with land certificates stated that the registration process and receiving the actual land plot were still complicated and seemingly impossible. General statistics of Khatlon province as well as study results show that few women own and manage farms.

---

45   "The economy of Tajikistan shall be based on various forms of ownership. The state shall guarantee freedom of economic activities, entrepreneurship, equality of rights, and the protection of all forms of ownership including private ownership [Article 12]....Everyone shall have the right to ownership and inheritance… [Article 32]." Government of Tajikistan. 1994. Constitution of the Republic of Tajikistan (amended 2003).

46   Analysis of multiple focus group discussions in the project area. October–November 2019.

3.   Women's Access to Financial Services

Rural female respondents of the in-depth interviews were limited in their access to financial services and, except for those who own a farm (with a legal status), most do not have bank accounts. Some families with labor migrants use local formal and informal individual operators to transfer money. The reasons for families using such operators are related to accessibility and swiftness of transfers. If such solutions are not available locally, people may travel to the village or town centers to make the transfers via a bank branch.

As an alternative, women from the same neighborhood (*mahalla*), groups of relatives, or wage workers often form groups "playing" the game of *casse* (*kassabozi*),[47] which helps to accumulate necessary funds for urgent expenses. Such informal financial instruments allow women to receive accumulated cash deposited by the group members and does not include any interest to be paid back. *Kassabozi* groups consist of 10 to 15 members who are mostly colleagues at work such as teachers or *mardikor* workers, and may be formed among several neighbors. The group appoints a cashier who keeps the cash collected in each selected period (usually monthly). In each round, the collected cash is given to a member who has priority needs, for instance, a wedding, or the purchase of bigger items such as furniture. The monthly amount collected per member may vary from TJS15 to TJS30 in the rural areas, depending on the level of earnings (e.g., teachers may contribute up to TJS50 per member per month) and the agreement among the group.

4.   Decision-Making

Economic empowerment is the ability to make decisions about economic resources. Even though a woman has earned wages, it does not necessarily mean that she can independently decide how her earnings will be used. Our survey included questions about decision-making. In most cases (over 85%), men do not see women as decision makers. In contrast, women perceive themselves as decision makers in the household and production process.[48] The women's perspective is understandable since a majority of them run households in the absence of male members and make their own decisions on a daily basis. Kitchen gardens are mainly in the domain of women who decide on type and quantity of crops, how much to spend on production inputs, and whether to consume or sell their produce. According to qualitative interviews, men would prefer to be seen as decision makers. They do not think women are capable or have sufficient knowledge to make decisions about the production process. Not surprisingly, men fully underline women's responsibility for cleaning, preparing food, health, elder and child care, and schooling.

5.   Access to Irrigation Services

Both I&D systems have particularities in their irrigation and agricultural production settings. These attributes are connected to environmental conditions, the landscape, and the I&D systems infrastructure. The I&D systems are complicated, unique structures built during the years of the former Soviet Union and are costly to repair and replace. While the irrigation systems remain functional,

---

47   During the study, *kassabozi* was mentioned as "playing" in a group, a *casse*, or being involved in depositing and money-lending activities among a group of individuals.

48   This response was the second most popular response.

many of the drainage systems have ceased to be effective due to lack of maintenance, resulting in high water tables and salinity. Alternative cheap and quick solutions, like diverting drainage water using subsurface pipes with filters to the main canal, could become a holistic and long-term approach for water management. WUA staff anticipate improvements with the transfer to the river basin management system. The River Basin Organization and its branches will be responsible for planning, monitoring use, protecting water infrastructure, and implementing the management of water resources in the basin zones.[49] ALRI branches (*Vodhoz*) will retain their roles as the implementing agency. A national information system is to be created for collecting, storing, processing, and distributing information for policy recommendation, prognosis, development programs, etc.

< Yovon irrigated system.
Deteriorated irrigation and drainage infrastructure leads to loss in agricultural lands in Yovon (photo by Nozilakhon Mukhamedova).

< Infrastructure deterioration.
WUA *mirob* adjusting the hydropost and checking the water levels in a highly land slide prone area in Yovon (photo by Nozilakhon Mukhamedova).

---

49   Information is based on the draft of the new Water Code of Tajikistan, December 2019.

5.1.   Households

Households in the project area usually occupy three to four buildings: living area and bedrooms, kitchen building, and a place for animals. Gravity or pumped water from the main irrigation canals is supplied through earth ditches and is used for multiple purposes. In both I&Ds, traditional domestic water systems also include water reservoirs or wells (*houvuz*) that meet the basic household needs for drinking, washing, cooking, and bathing. Households situated in the tail end or located far from the irrigation system may not be served by water service providers. The multiple uses of water by female household members indicate the importance of uninterrupted access to safe and adequate water for women.

Female members of the households contribute significant time and effort to family care and domestic work. Such specialization is considered conventionally as the female's exclusive responsibility and obligation to the family. Women are responsible for water allocation and management within the household and its use for household chores and to maintain hygiene. It is quite common in Tajikistan for women to be in charge of collecting or carrying water from water sources. This work is time-consuming and leaves little room for education or professional purposes, as well as recreational activities, which in turn undermines a woman's more active role in societal pursuits.

Households also use these water systems to support a wide range of productive activities such as small-scale growing of crops and vegetables, raising livestock, and for informal small business operations. The prioritization of the water management organization on productive uses of water by registered farmers neglects the kitchen garden operators' claims to water and the value of their labor in the domestic realm, especially if the farmers and kitchen gardeners use the same water source.

Households in the study area usually make payments for irrigation services to the local government representatives. However, WUAs do not always respond when households are not receiving enough or any water. Therefore, some households stop paying fees to WUAs and organize water provision on their own. In Jayhun district, households in one village use pumped groundwater for drinking, which is metered.

Households of Huroson district in the tail end of the Right Tributary I&D system cannot grow fruit trees or plant crops apart from fodder since the drainage water is too close to the surface. Kitchen gardens also illustrate the issue of unequal water distribution between farmers and households. This is especially evident from WUA irrigation services provided through pump stations, which work only from March to mid-October, leaving people to find ways to reserve, collect, or buy water during the rest of the year, affecting both drinking and subsistence agricultural production needs (Box 1).

Payments for irrigation service does not cover all the necessary O&M works such as cleaning the canals. Households and farmers are involved in joint voluntary community work to clean the irrigation and drainage canals. Although male respondents constantly said that such work is not done by females, many women are involved in the work. Some households have heard of WUAs, but were unsure about their function, membership, or responsibilities. Males collect the WUA fees from the households, where most payees are females who make the garden production decisions. To address the mismatch between patriarchal practices and migration, women should assume fee collection jobs.

5.2   Irrigation of *Dehkan* Farms

When a *dehkan* farm registers with local authorities, it has to conclude an agreement with a WUA or, in areas not covered by a WUA, with an ALRI branch. Division and provision of water by WUA *mirob* workers to their members is based on pre-agreed terms. If *mirob* workers do not have the capacity, time, or finance to cover all the users, the farmers sometimes open the gates themselves. Usually, farmers irrigate land manually. Only 11% use drip irrigation (Box 2). Farmers often use the service of a *mirob*, who can serve several farmers and receives TJS100 a day.

Most (80%) male respondents consider that there are no irrigation problems. However, more than 50% of female respondents state problems such as: lack of water, bad condition of canals, users at the head of the canal taking most of the water, natural drought, and unsatisfactory work by WUAs. (Figure 5).

Box 1: Drinking Water— A Vital Resource

"Winter time is a bit complicated for many households in terms of water supply. With the end of the farming season, all the irrigation systems and pumping stations stop or work partially." The female household member, who lives in Yovon in the Right Tributary system, said that for about 2 months in winter their water supply gets cut off. "When the pumping station is closed we do not have any water. During the winter...they let out drainage water and close the canal water [and it is] only in March [that] the normal water comes back. During the winter we drink rain and melt snow. We also bring water in buckets or 20-liter containers, but it is around 1 kilometer. We have a container where we harvest such water. It would be good to have water also in winter. However, we still consider to live in good conditions as there are other areas where people have to buy water."

Source: Interview with a female household member. October 2019. Yovon, Right Tributary Irrigation and Drainage system. October 2019. Yovon, Right Tributary Irrigation and Drainage system.

Box 2: Taking Action for Better Service

"During the irrigation season, the water was disappearing, the *mirob* [workers] did not pay any attention and we stopped paying the irrigation service fee to WUA. Instead, we installed our own pump, collecting TJS80 from each user for a pump and also installing a water meter into the pump. This way, each person knew what he should pay."

Source: Interview with a member of a group of households in the Kumsangir canal area. October 2019.

Figure 5: Main Issues Related to Irrigation

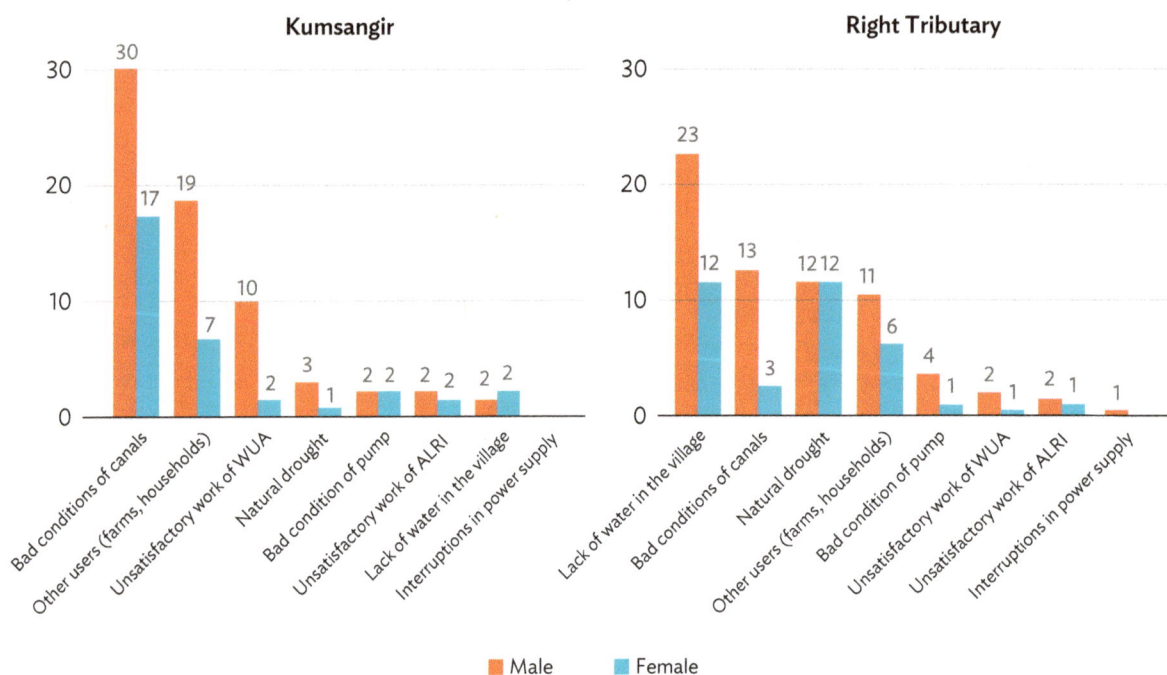

ALRI = Agency for Land Reclamation and Irrigation, WUA = water users association.
Source: Data analysis based on individual household interviews (n = 323). November 2019.

6.   Patterns of Agricultural Production

6.1.   Kitchen Gardens and Cropping Patterns

The kitchen garden occupies usually a third or a quarter of the residential land. Households of the Right Tributary plant up to three crops during a season. The crop combinations include potato–corn–greens, potato–maize–potato, and carrots–cabbage–potato. In Jayhun district, kitchen garden production is more than subsistence farming. Jayhun is famous for its vegetables and lemons. Since the current government allowed greenhouses, farmers started building them on both *dehkan* farms and household plots. Annually, a farmer with around 0.5 ha of land and approximately 400 lemon trees can yield up to 20 tons by the third or fourth year.[50] Lemons are exported to Afghanistan, Kazakhstan, and the Russian Federation.

Based on qualitative interviews, elder women decide what and when to plant in their kitchen gardens and distribute the work among daughters or daughters-in-law. While the men might help with the land preparation, they are rarely involved in kitchen gardening. Households may sell any surplus harvest to neighbors or in village markets. Women themselves sell and control the income of their kitchen gardens, except in cases of greenhouses in Kumsangir, where men often take the lead of profitable businesses such as lemon growing.

---

[50]   During the period of the interview (October–November 2019) the lemons were abundant in the local market and sold for TJS5 per kilogram. Intermediary buyers offered TJS4,000–TJS4,500 per ton for lemons grown in Kumsangir.

Household subsistence farming. Kitchen gardens run by female household members in Kumsangir (photo by the author).

Small-scale home businesses. Home-based lemon growing greenhouse in Kumsangir (photo by Aziza Abdullojonova).

6.2.   Popular Crops

Survey respondents grew several types of crops (Figure 6). Nearly half the male respondents specialize in cotton, followed by vegetables and wheat. Females in both canals identified vegetables as their primary crop followed by wheat and cotton. Melon and fruit production is low.

Figure 6: Crops Planted by Male and Female Respondents
(number of respondents)

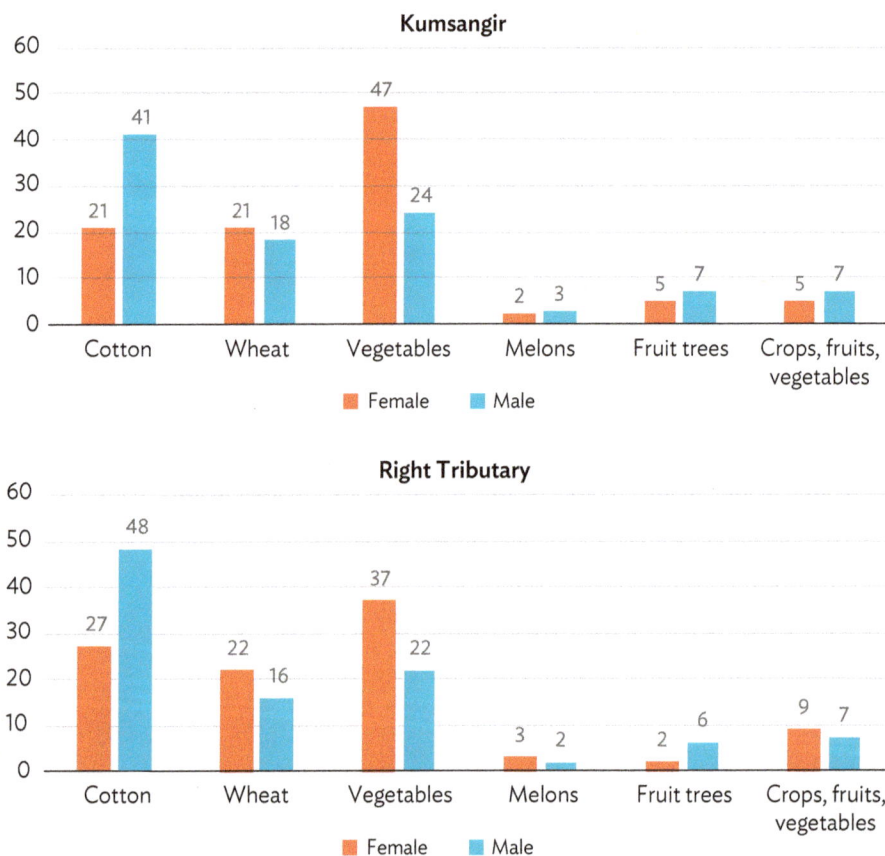

Source: Data analysis based on individual household interviews (n = 323). November 2019.

Cropping areas vary. Cotton and wheat cover most of the total crop area used by the 323 respondents. More than 47% of total land is for cotton production and 20% is for wheat. Alfalfa and maize are grown on 13% of land. Potatoes, onion, melons, vegetables, beans, and fruits are grown on the rest of the cropland. The most important crops are cotton, wheat, and potatoes because of their profitability. All producers in the Right Tributary mentioned that they cultivate cash crops that provide access to hard currency.

6.3.   Agricultural Productivity

Productivity is determined by such factors as soil quality, land accessibility, capital and labor, and farmers' characteristics.[51] Productivity analysis shows that men perform better in production of wheat, maize and other cereals, potatoes, onions, melons, orchards, and beans. Female farmers are more productive as cotton growers in Kumsangir and as vegetable growers in the Right Tributary.

For fertilizers, farmers commonly use nitrogen, phosphorus, potassium, and pesticides; diesel fuel is used for machinery. The main source of all inputs is specialized private suppliers. Half the respondents obtain seeds for wheat and vegetables from private suppliers. Besides private stores, cotton producers buy seeds from state companies and agroprocessing enterprises on a contract basis. Issues in securing inputs include lack of finance, high prices, poor availability in markets, and transportation.

Half the respondents said the main factor in productivity is supply of irrigation water. Availability of good quality seeds is also important. Access to information, credit, and machinery services are also important factors particularly for females.

6.4.   Challenges to Women's Productivity

6.4.1. Climate Variability

Climatic patterns in Tajikistan are becoming more unpredictable, resulting in erosion problems on sloping terraces, mudslides, flooding, and increased water demand in summer. Such impacts are making planting decisions an increasingly riskier undertaking. Female farmers and female-headed households are frequently among the most vulnerable in rural areas, and often have very limited capacity to cope with or recover from weather-related losses. Respondents from Yovon (23%), and Kumsangir (18%) reported climate change as a serious issue.

6.4.2 Access to Information and Farming Skills

Females and males in the project area have differences in access to information and training opportunities. Among respondents, 109 males (34%) and 41 females (13%) had participated in agricultural training. Training was delivered mainly by authorities or farmers' unions, international organizations, NGOs, and extension service firms.

The sex-disaggregated data show that "laser leveling" and "quality and export" training had zero female involvement but in other topics, most female respondents have been trained in aspects of vegetable, fruit, and cereal production (Figure 7).

---

51   In the present case, productivity is calculated through yield over land size.

Figure 7: Application of Knowledge and Skills Gained from Training by Male and Female Respondents
(%)

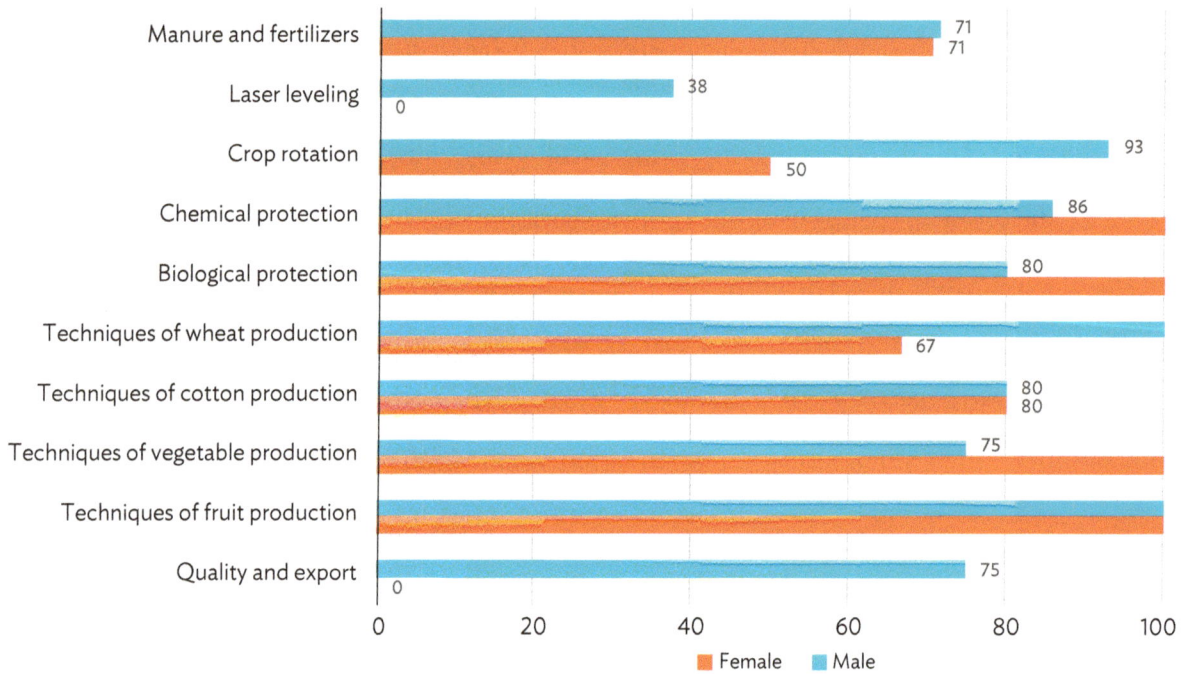

Source: Data analysis based on individual household interviews (n = 323). November 2019.

6.4.3 Access to Agricultural Services

Agricultural services in the study area are fully or partially provided by the state; other providers include international institutions and nongovernment entities, predominantly WUAs, and to a lesser extent, farmer groups, and microcredit institutions.

More than half (56%) of female respondents in Kumsangir and 39% of those in the Right Tributary system were members of WUAs. Membership in WUAs is tied to the legal status of a farmer and according to WUA by-laws, only members are eligible to receive irrigation services. This excludes women nonmembers operating kitchen gardens and contributes to their time poverty and drudgery, influencing in turn the household's food security and reducing their participation in economic activities. Membership in farmer groups[52] followed the same trend, 14% females in Kumsangir and 7% females in the Right Tributary I&D. Receipt of microcredit services was very low among female respondents, 4% in the Right Tributary, and none in Kumsangir, which could reflect availability of financial institutions and services in the respective areas. Membership in cooperatives was absent in the study I&Ds.

---

[52] Established as Farmer Advisory Services in Tajikistan under a USAID project to improve the livelihoods of small farmers by supporting an agricultural extension and advisory system in Khatlon province and providing support for agrarian reform.

D. Labor Arrangements and Feminization of Agriculture

Tajik women have three roles in agriculture: agricultural wage or daily workers, managing kitchen gardens, and *dehkan* farmers. These roles may overlap, especially the role of women as caretakers, small agricultural producers within their homes, and their visible roles as wage or daily workers that is often not recognized in official statistical reports. The differences between these roles comprise decision-making levels, flexibility of roles, allocation of time, payments for work, and recognition of the role in official accounting and labor system. Women's management of kitchen gardens is usually unpaid but requires consistent allocation of time and is related to subsistence farming. Wage or daily work is flexible, optional, and allows women to earn for themselves and their families.

1. Changes in Labor Structure

The current agricultural labor setting is characterized by labor outmigration, greater female participation in agricultural activities, and shortage of formal jobs.[53] Reforms and sociodemographic changes in the labor market after the fall of the former Soviet Union caused significant movement in agricultural labor arrangements. Formerly, agricultural workers of the collective and state farms had stable work and constant salary. Gender relations were regulated by involving women in production on a similar basis as men. The representation of men and women in state and collective farms was almost equal. Both genders participated equally in many activities, such as cotton production. Although women were expected to work in the farm, whether paid or unpaid, the household income was guaranteed by salaries and was enough to sustain the family. Organization of social security (maternity leave, decent pensions) and functioning public institutions like kindergartens, provided opportunities to work.

After the farm restructuring, not everyone could become farmers or continue as permanent farm workers. For the rural labor force, restructuring of collective and state farms meant a shift to other available work or new jobs in other sectors or outside Tajikistan. At the same time, new farming units, such as *dehkan* farms, gradually increased in number and production levels, which created demand for workers. These conditions drove former Soviet Union farm workers (*kolkhozchi*) and the new generation of the rural population into agricultural work offered by *dehkan* farms. However, individual farms could not offer the same labor arrangements, and the state could not maintain the social security system over the period of structural reforms. *Dehkan* farms started hiring workers through informal channels to avoid the 38% taxation[54] for hiring low-cost seasonal workers (Box 3).

> **Box 3: Making Seasonal Jobs Attractive and Accounted**
>
> "Our legislation should adapt to accommodate seasonal workers. I am a farmer and I always need additional labor in my farm. Let's say if I pay TJS1,000; if...I am taxed then I would have to pay 38% on top to the tax office for hiring *mardikor* workers. The government wants us to employ them, but it's too costly. We need to make changes in the current legislation. Perhaps it is a small detail, but it's negatively influencing on attracting labor into agriculture. That is why we do all with our own efforts. The tax office knows that larger plot where vegetables are grown would require lot of seasonal workers and demand on paying taxes."
>
> Source: Interview with a male *dehkan* farmer. October 2019. Buston-K WUA, Jayhun district, Kumsangir Irrigation and Drainage system.

---

[53] Formal jobs are assumed to provide workbooks, social benefits, paying taxes, etc.
[54] Personal interviews with *dehkan* farmers. October to November 2019. Jayhun district.

Female seasonal workers filled the male labor gap in the agriculture sector. In some cases, women replaced men in male-dominated positions such as in tilling land, harvesting, and irrigation, as well as accounting and fee collection in WUAs. Female participation was driven by lack of household income and/or unreliability of remittances.

It is difficult to measure the feminization phenomenon in terms of increased number of workers as it would require comparing workforce figures prior and during the assumed feminization period. The informality of job arrangements means that farmers are less likely to reveal information about all labor involved in their agricultural production activities. This study found that more than 80% of women were involved in predominantly informal agricultural work. Thus, they were not registered as official workers and were ineligible for social support or pension provisions.

^ **Land tenure security.** Cotton picking by female seasonal workers in the Right Tributary (photo by Nozilakhon Mukhamedova).

Female agricultural wage workers are paid based on the type of activity, crop, and workload. There are three payment types: daily rates, piece rates, and payments per type of crop and activity (Table 6). Daily rates vary by district and time of year. In Jayhun district, the rates for daily wage workers in November were equivalent to $500–$800 annually.

Table 6: Payments to Seasonal Workers for Major Agricultural Activities and Crops

| Crop Type | Tilling | | Weeding | | Trimming or Rooting Out | | Harvesting | |
	Frequency per Season	Daily Wage (TJS)	Frequency per Season	Daily Wage (TJS)	Frequency per Season	Daily Wage (TJS)	Frequency per Season	Daily Wage (TJS)
Cotton	2	50–100	2	40	1	60	3	45–80
Carrots	2	50–100	2	40–60		50–70	2	60–70
Onions	2	50–100	2	40–60		50–70	1	50–70
Fruits[a]								30

TJS = somoni.
[a] Persimmon, plums, and cherries.
Source: In-depth interviews and focus group discussions with *dehkan* farms and *mardikor* groups. October–November 2019.

Structured individual interviews included multiple choice questions to reveal opinions of both male and female respondents about major agricultural positions held either by men or women. Answer choices for these questions were built on observations and interviews of exploratory visits to the project sites (Table 7).

Table 7: Categorization and Description of Agricultural Occupations

Agricultural Occupation	Description
Kolkhozchi (collective farm worker)	A term carried from the past and translated as Soviet collective farm worker. Presently, this refers to hired workers mainly involved in cotton production[a]
Mardikor/yordamchi (day laborer or helper)	Hired daily workers
Household work, kitchen garden	Work connected with agricultural production in the household plots
Sahmdor (land shareholder)	Shareholder and registered employee of a *dehkan* farm[b]
Mirob (water master)	Water masters involved with irrigation services
Manager, agronomist, or brigadier	Person who can make agricultural management or production decisions

[a] In Northern Tajikistan the term "*hectarchi*" is the equivalent of *kolkhozchi*.
[b] N. Mukhamedova and K. Wegerich 2018. The Feminization of Agriculture in Post-Soviet [Union] Tajikistan. *Journal of Rural Studies*. 57. January. p.135.
Source: In-depth interviews and focus group discussions with *dehkan* farms and *mardikor* groups. October–November 2019.

2.   Women's Labor Participation

Hired workers (*kolkhozchi*) and kitchen gardening agriculture positions were the most common among women (Figure 8). Around 35% of the female respondents and 27% of the male respondents recognized women taking hired daily workers (*mardikor* or *yordamchi*) and land shareholder positions. Females dominated all positions even though the popular opinion among the respondents is that some are in the male domain.[55] Respondents felt there was a deficit of workers in such male positions. For example, fertilizer inputting was attributed as men's work by both males and females. But a lack of male workers to perform this activity necessitated finding different solutions. Since the local population considers fertilizers to be toxic to female health and affect pregnancy,[56] this job is now mostly performed by youth (aged 14–20 years). Women handle kitchen garden irrigation themselves. Some female farmers use the service of a *mirob*.

Perceptions of female and male respondents on characteristics of female jobs were similar; two-thirds of the interviewees chose the jobs demanding low physical power. This result follows the customary gender-based division of labor in Tajikistan and the patriarchal views of women's role as a caretaker and performing domestic work. Other job criteria for females included those that do not demand any professional background, are appropriate with the religion, or require detail and patience. About 30% felt that there was no criterion; women can and do every job.

The "most important" or "important" position was the *kolkhozchi*, which can be attributed to previous work in collectives and state farms or current jobs related to cotton production that could be tagged with the *kolkhozchi* job title. In both I&Ds, registered employees and hired daily workers (*mardikor*) may

Figure 8: Types of Agricultural Work Currently Performed by Men and Women
(%)

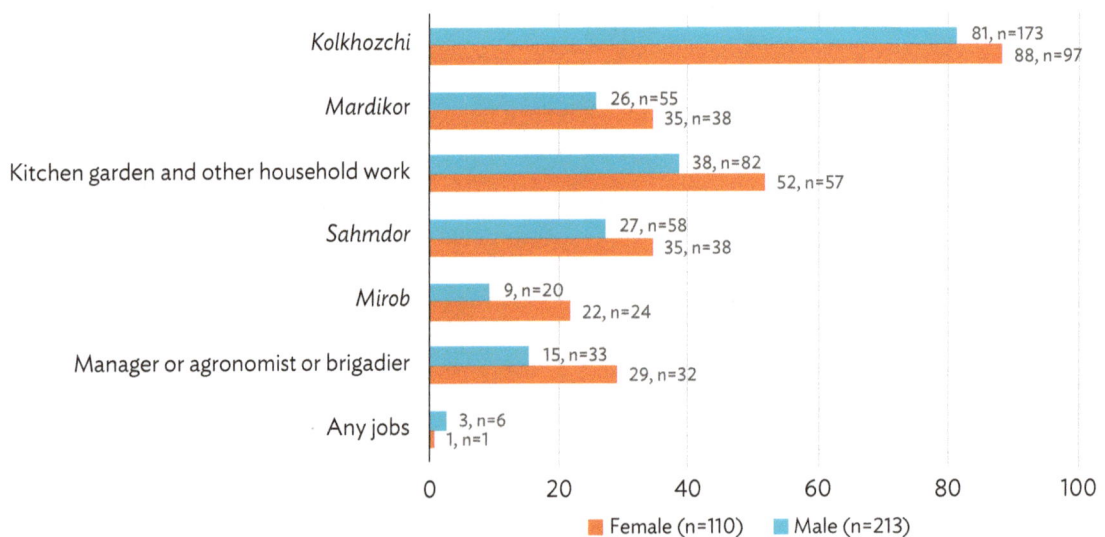

Source: Data analysis based on individual household interviews (n = 323). November 2019.

---

[55]   Results from opinions on male positions: *mirob*: 41% male, 55% female and manager, agronomist, or *mardikor* leader: 29% male, 43% female.
[56]   Personal interviews with female in various positions, Khatlon province, October–November 2020.

be involved in cotton production activities.[57] Farmers growing cotton usually have registered employees or other permanent workers involved in all phases of production and pay a salary and additional in-kind products grown by the farm. *Mardikor* workers are attracted to cotton harvesting since they are paid for their effort (a fee per kilogram of harvested cotton), but also because of possibly receiving additional in-kind payments like cotton sticks.[58] Many female respondents reported that they have been involved in such work for several years and often were re-invited by the same *dehkan* farmers.[59]

3.   *Mardikor* Groups

3.1.   Why Women Join *Mardikor* Teams

Mardikor workers mostly perform agricultural work, and 98% of them (in Northern Tajikistan) are female.[60] Box 4 presents insights about the reasons for doing wage work usually performed jointly in groups or "brigades."

Mardikor workers are mostly under the age of 50; the youngest are 13–15 years old. They do not have formal, written contracts, are not registered as farm workers, and cannot receive social benefits. Oral agreements are stated weekly or daily. Females are involved for 9–10 months in seasonal, intermittent *mardikor* jobs for an average of 12–15 days per month.

Box 4: Reasons to Do Wage Work

"My husband left to work [in the Russian Federation] 2 years ago. He sends remittances of TJS300–TJS400 (about $35) every 2–3 months. This is enough as we also plant carrots, potatoes, onions in our kitchen garden. We do not have access to a land (land rights). But I also have to work to support my children. Although, we live with our extended families, I have to earn money too. Monthly costs for food and clothing or school items for our family with two children is approximately TJS1,200 (about $123). I [have been] working as *mardikor* [for] 5 to 7 years. During the season of onion harvesting, I can earn up to TJS500, however that's only during a certain period. For works like hoeing, weeding one might earn only TJS5 to TJS6 per day.

We have to do *mardikor* works. In one household, you can find three to four families and they all have to be sustained. We have to work [to sustain them], life obliges. We have to work [for the] sake of our children."

Source: Interview with a *mardikor* worker. November 2019. Jayhun, Kumsangir Irrigation and Drainage system.

---

57   Personal interviews with *mardikor* and *sahmdor* workers in October–November 2019, Khatlon province, Tajikistan.
58   Cotton sticks are mainly used as firewood for baking bread and other food preparation.
59   In Sughd Province there was a specific position, *hectarchi* or *kolkhozchi*, which is a female performing various works together with neighbors or with other family members during cotton growing and harvesting. *Hectarchi* usually is an informal farm worker who can inquire or be asked by the farm managers about taking a small cotton plot. Usually, when the salary or cotton harvest is low, the farmer pays the wage in-kind and allows the *hectarchi* to collect the cottons sticks that serve as firewood for their families.
60   "The *mardikor* position, which could likewise be considered undesirable in terms of social security compared to other agricultural employment groups, is actually considered the most preferable and provides the highest income. Therefore, this position would be the best for covering daily household expenses. *Mardikor* workers take the least secure position, which is sensitive both to the seasonal availability of jobs and the harvest yields. *Mardikor* workers also work under pre-agreed outputs in days and hectares and receive daily payments (*kun-bai*). Since the *mardikor* position is the best-paid job, it has also contributed to a shift in bargaining power within the household. Hence, through the new position as *mardikor*, women receive recognition as income earners and providers for the family (footnote 3).

3.2.   Types of *Mardikor* Work

Agricultural jobs are mainly connected with weeding, tilling, and harvesting. Activities start early March with tilling and end in November with collecting firewood from the cotton fields. Female *mardikor* workers combine different livelihood sources and may fully or partially depend on what they earn from their seasonal or daily activities (Box 5). Agribusiness expansion and the inflow of investments from neighboring countries have brought new activities and workers go through training offered by the farmers; women help to set out drip irrigation tubes and laying out plastic material on top of planting rows.[61] Assigning activities with new technologies to women is further evidence of increased female participation.

Box 5: Every Wage Worker Should Be Accounted

"Each *mardikor* leader has her own way of setting terms of the agreement with farmers. I register all the women in the group for [a] certain activity. I negotiate with [the] *dehkan* farmer the payment based on the furrow count. One furrow is TJS4 to TJS5; women usually can cover 8-10 furrows, so I would check my registry book every day to see who worked and how much and pay out the group members TJS40–TJS50. In case of cotton, it is weighed at the end of each day and *mardikor* workers are paid for 10 days of work. A kilogram of cotton is TJS0.50–TJS0.60. Harvesting [a 50-kilogram sack of] carrots: TJS2.50. Working as a *mardikor*, one can earn TJS300 up to TJS600 a week.

As a leader, I can earn as a *mardikor* and as a *mardikor* leader at the same time. If a *mardikor* works well on the land and if you are also a *mardikor* leader you could earn from TJS300 up to TJS600 a week. During the onion season, I can earn up to TJS2,000 a month and usually it's TJS1,500 a month."

Source: Interview with a *mardikor* leader. November 2019. Jayhun, Kumsangir Kumsangir Irrigation and Drainage system.

Working on crops such as carrots or potatoes is cumbersome and requires considerable effort. Harvesting carrots also includes cleaning the carrots, cutting out the green shoots, and packing into 50-kilogram sacks. *Mardikor* workers have to buy their own tools. For example, they need wide scissors (TJS9), shovels (TJS15), and gloves. Female agricultural workers earn $400–$700 per season.

The easiest job, according to the group of *mardikor* workers interviewed, is cotton picking. Usually, cotton pickers can collect 90–150 kilograms per day. They get paid TJS0.50–TJS0.80. Apart from harvesting, females can get, on average, TJS150 a month (but in companies from the People's Republic of China, TJS800–TJS900/month) for cotton activities and in-kind payment of cotton sticks used for firewood. However, such remuneration is very low, and if a female does not receive any remittances and relies only on herself, it is very difficult to live and survive.

---

61   Drip irrigations and plastic covering was used on several farmlands rented by companies from the People's Republic of China.

^ **Daily workers.** *Mardikor* team of female harvesting carrots on the *dehkan* farm field in Kumsangir (photo by Nozilakhon Mukhamedova).

3.3.   *Mardikor* Conditions

Mardikor women almost always work in brigades or teams of 30–50 people. In each community, there is commonly at least one and up to four *mardikor* groups. In the summer, working hours are 8:00 a.m. to 6:00 p.m. and in winter, 9:00 a.m. to 4:00 p.m. Conditions can be harsh, especially during the summer heat (above 50°C) and the cold in November. They mostly work for large *dehkan* farms that also have land shareholders.

Females in *mardikor* groups are usually not registered as agricultural workers and do not have workbooks. They expect to get the minimum official payments guaranteed by the state when they are on pension. While working, they leave their children with their parents or other older members in their household or bring them to the field, where they help with the work.

Females usually spend their earnings on school and other expenses for their children[62] and their own needs. The majority of children, besides school, help their mothers and older sisters with agricultural work. Children go to the fields after school, especially during harvest season. With male outmigration, many restrictions for women have been eroding, allowing the possibility to work outside the village

---

62   "We have to pay TJS13 per child per month. I have two children going to school, plus I have to buy their clothes and school items." Interview with a female *mardikor*. November 2019. Buston K WUA, Jayhun district, Kumsangir.

and travel.[63] Female agricultural workers recognize traveling outside their communities and villages as a necessity and an opportunity to get more work. Female *mardikor* workers travel in groups and brigade leaders assume responsibility and protect the group through networks of farmers in other villages.

3.4.   *Mardikor* Leaders

Leaders of the female work groups are experienced negotiators and team managers. They receive additional remuneration for their organizational efforts and management skills. They also work together with other group participants and at the same time retain control over their work. One important skill is a leader's ability to negotiate with male farmers hiring her *mardikor* team. The wages are set by group leaders who explore current rates for various agricultural jobs and negotiate in advance with farmers the terms and amount of payment for each *mardikor* female in the group. Still, *mardikor* group leaders do not have enough power and capacity to increase the prices set by farmers in the market, which are close to the minimum wage in the country.

Mardikor leaders (Box 6) receive additional remuneration from the farmers, usually TJS30/day. They are pioneers in changing conventional restrictions and stigma for women of working for various individual farmers and in various villages. Although the concept of "brigada" (*mardikor* groups) existed in Soviet *kolkhozes* and *sovkhozes*, the work relations were based on formal hiring and membership in collective and state farms. After the farm restructuring and individualization of farms, labor relations changed from centralized hiring to hiring separate individuals or groups (mostly female) by private farmers (mostly male). Female *mardikor* leaders made efforts to negotiate and explain to the husbands and other family members the arrangements for female wage workers and the advantages of their jobs. *Mardikor* leaders also worked to create credibility and good reputations around their teams to change entrenched patriarchal opinions to allow females to take such jobs. *Mardikor* group leaders continue to empower females and are keen on building constant work opportunities and networks of farmer–employers to provide security and stability over time (Appendix 5).

Mobile phones have become essential tools of communication for females, who contact the *mardikor* leader themselves to learn about work availability. It is practical for farmers to have connections to at least three *mardikor* leaders and inform them about the type of crop, activity, hectares of land, and remuneration. *Mardikor* leaders then figure out how many people they need and for how many days. They check for available people on their lists and inform them about the job opportunity.

---

63   Interview with *mardikor* women. October, 2019. Huroson district, Right Tributary Irrigation and Drainage system.

Box 6: How I Became a Brigade Leader

"How did I start to be a *mardikor* group leader? It was not easy in the beginning. I heard about two women who organized such work in other districts. Rumors about such work spread in our village and women got interested. This was in 1994 or 1995. I decided to organize a group. I asked permission from my husband, my father, and my brothers at that time. They agreed. In the beginning, we had 15 women working and earning money. Our brigade started expanding. I have phone numbers of all the female *mardikor* [workers] and the farmers. I have to be responsible for the work time, money received, and security."

Interview with a *mardikor* group leader. October 2019. Buston K WUA, Jayhun district, Kumsangir Irrigation and Drainage system.

Chapter IV

CONCLUSIONS AND RECOMMENDATIONS

Agriculture remains Tajikistan's dominant economic activity. Over 70% of the population living in rural areas and many workers, virtually all males from rural areas, have outmigrated in search of better opportunities. This has forced women to assume more responsibilities in the household as well as seek other income opportunities. Since remittances may not be stable or reliable, females have had to supplement their household income through various, mostly informal jobs. As a result, they have become the dominant labor force in agriculture, with daily female workers making up over 80% of agricultural labor. The "feminization of agriculture" that took place in Tajikistan and other transition countries, includes women being employed in traditionally male-dominated occupations. Agricultural activities that used to be performed equally by males and females, such as cotton growing or tilling or harvesting vegetables and fruits, are now performed mainly by women.

Although the male respondents do not openly oppose the rising number of females in conventionally male-dominated positions such as *mirob* workers, irrigation fee collectors, *dehkan* farm managers, and managing positions in WUAs or ALRI, they consider these as "not a [female's] job." Females face gender disparities and inequalities not experienced by males. Females as decision makers are in the minority or absent among *dehkan* farm managers and therefore are not visible members of WUAs. Clearly, the expanded role of women needs recognition and support in practice as well as in legislation.

Based on the findings of the study and strategic priorities of ADB's Strategy 2030, the following recommendations focus on women's economic empowerment and reduction of their time poverty. The target groups for future gender interventions and projects are females whose husbands have migrated and joined already existing female *mardikor* teams. They are involved in lower-paid agricultural jobs, have restricted access to production inputs and decision-making, and/or operate kitchen gardens. Recommendations target four key areas of intervention: (i) land tenure security; (ii) access to inputs, knowledge, and finance; (iii) gender mainstreaming in water governing and managing institutions; and (iv) labor participation in agricultural positions.

A. Land Tenure Security

Legally, males and females in Tajikistan have equal access to land. However, in reality, females are at a disadvantage with the complicated land registration process and lack of legal knowledge and financial support. Female respondents tend to have smaller land plots than males and often lack the self-confidence to run a farm. Lack of access to secure land restricts females in their productive activities and income generation. However, even if females express the will and confidence to manage their own farm and to take a risk, they still face obstacles in accessing the land. In legal terms, becoming a *dehkan* provides access to both land and water. Bureaucratic, social, financial, and knowledge-related gaps hinder females from managing a *dehkan* farm. Overcoming these barriers would guarantee equal tenure security, improved productivity, and involvement of women in decision-making processes.

Recommendations

(i) **Access to legal advice.** Provide legal advice for women who already have land shares but could not yet secure land use rights and practical support to women to complete the land registration process, register and establish a farm, and be able to manage and maintain their land.

(ii) **Access to land use rights**. Provide land for groups of females, especially *mardikor* groups, who do not have shares of land but would like to access land and get involved in crop planting, and be registered as cooperatives. These steps would improve equality in access to land and allow females to have a greater stake and say in irrigation issues. Financial instruments such as microcredit loans along with relevant technologies and training should be included.

(iii) **Access to connectivity.** Provide consultation services through mobile phones, which are owned by the majority of females. Internet is expensive and not widespread, thus a call center with consultative services on financial products, money transfers, and legal and other agriculture-related advice could be established under local government offices, WUAs, or as a business.

B. Access to Inputs, Knowledge, and Finance

Although females are in a minority among farmers who own and manage full agricultural value chains, they show equal or even better performance than their male counterparts. Female farmers are less involved in cotton production, but highly specialized in growing vegetables. There is a big gender gap between female and male participation in the value chains. Difficulties for females in obtaining production inputs are attributed to the lack of access to finance, the high prices of farm inputs, the poor availability of inputs in the markets, and the absence of extension services and transportation. Female farmers have restricted access to markets outside their villages and lack options and finance to invest in storing, packaging, and/or processing their produce. They also lack access to knowledge and skills; no formal or nonformal vocational training or services are available for females.

Female farmers struggle with access to financial services and cannot always reach banking service points or money transfer machines. Although many women have mobile phones, these are often not suitable for banking operations. Internet connections are costly and many banks do not offer online banking services. Such financial hindrances also complicate the receipt of remittances sent by their migrant relatives. Most rural women do not have bank accounts as they are not formally registered as agricultural workers.

Recommendations

(i) **Access to quality inputs.** Improve females' access to quality inputs (seeds, seedlings, and saplings), technologies, and machinery to increase yields and production value. This is important because diversification to high-value crops entails access to quality and climate-resilient seeds and improved practices and technologies in irrigation. Access to high-value seeds could be channeled through national seed banks or private sellers. Another localized option for seeds is establishing a network of male and female seed growers and creating a seed bank especially for local and regional high-value crop and tree varieties.

(ii) **Access to finance.** Improve access to financial products to provide possibilities to invest in farming and agriculture-related businesses. Preferential loans specifically for women can target leasing or purchase of machinery, technologies, seeds, or fertilizers. Access to banking services for simple money transfers should be provided for women who are distant from most bank offices and cannot travel far.

(iii) **Access to extension services.** Improve access to knowledge and information through targeted training for rural women to help them diversify into high-value crops. Training can be tailored to provide farming and business skills as well as legal support for women. Long-term interventions include establishing vocational schools or small agricultural production or processing businesses; and working with other stakeholders to produce video materials with the participation of successful actors (male and female) along the value chain. Video presentations can be distributed to rural populations to WUAs, local governments, and via local TV channels. Priority in training should be given to female farmers whose husbands have migrated; female *sahmdor*, agricultural workers who cannot receive their physical land plots owned according to their land use rights (land certificates); and rural females without shares as well as wage workers belonging to *mardikor* groups who provide agricultural services and plan to become farmers.

C. Water Governing and Managing Institutions

Data provided by ALRI have limited information on socioeconomic and institutional aspects of water management. Requests for information could take up to 6 months to be answered, as the requests should be made in writing to multiple agencies and waiting for their answers.

Around 20% of female respondents among *dehkan* farmers experienced verbal or physical abuse when asking for water. The issue, however, is unknown to male respondents and WUA staff. Further understanding of the issue and finding solutions should be a priority and included in the gender strategies of ALRI and WUAs. The first step may be to design gender sensitization campaigns around this issue.

Females are not always accepted to managerial positions where they are equally eligible as specialists or managers in ALRI. Key respondents expressed their wishes about female hydro technicians and ameliorators to be recognized, made visible, and to have equal conditions for applying to management positions and receiving the same compensation as male colleagues.

The ALRI gender group needs to expand its activities, plan for gender mainstreaming indicators, and increase the supply of skilled and professional females through networking with universities and vocational training programs. Introduction of formal institutional policies and practices that are purposefully designed to create a conducive environment for the recruitment, retention, and advancement of qualified females in ALRI are also needed.

Recommendations

Agency for Land Reclamation and Irrigation

(i) **Capacity enhancement.** Build the capacity of ALRI for designing and implementing a gender strategy with an activity plan for the ALRI gender group, including activities for engaging with male staff on gender issues. A basic minimum wage along with equal wages for males and females should be an objective of the strategy.

(ii) **Data management.** Advise and support ALRI and its branches in creating a common gender-disaggregated database from which they can derive useful analytical and practical results of their work.

(iii) **Policy framework.** Introduce gender-sensitive staffing in ALRI and its branches, including formal institutional policies and practices purposefully designed to create a conducive environment for the recruitment, retention, and advancement of qualified females.

(iv) **Gender awareness and sensitivity.** Raise gender awareness and conduct annual mandatory training on the attributes and expectations of a gender-friendly work environment for management and staff. This can shift mindsets and organizational culture over time, particularly when policies and expectations are enforced and tied to performance and reward systems for all employees. As females are not always accepted to managerial positions where they are equally eligible, incentives to employ more female staff can be included.

(v) **Education and skilling.** Increase the supply of skilled and professional females through networking with universities and vocational training programs.

(vi) **Enabling environment.** Develop peer networks and organize communities of practice designed specifically for female water practitioners that can help ease feelings of isolation and stigma often experienced by pioneering women who are the first to enter male-dominated occupations.

(vii) **Shift in design approach.** Apply a user-centric design approach to water supply systems that incorporates end users' wants, concerns, and cultural contexts. For this approach, ALRI should ensure it is women friendly, and women users are consulted in the design.

Water Users Associations

(i) **Data management.** Create a sex-disaggregated statistics database. General statistics as well as specific WUA-collected data are based only on the first crop of the season. However, where cotton and wheat are prevalent, data on the second cropping play a huge role in understanding the diversification and intensification of agriculture and evaluation of nutritional value of crops. The second crop harvest also serves as an in-kind payment for the wage workers or members of the farm.

(ii) **Access to opportunities.** Form all-female subcommittee(s) through which members can be gradually and systematically be trained in leadership skills.

(iii) **Access to infrastructure.** Ensure equal provision of water infrastructure and access to irrigation water. Infrastructure such as gates, pipes, and filters should be designed to reflect the needs and preferences of females as well as cultural and social village settings. Females should be introduced to new water management systems and technologies where they control the irrigation process and have transparent accounting with water service agencies.

D. Labor Participation and *Mardikor* Groups

Labor outmigration is a partial solution to rural unemployment of the male population in the study area. At the same time, new farming units—small- and medium-scale commercial farms—have increased in number, expanded their production activities, and created more demand for workers. Previously, collective and state farm employees and the new generation of female agricultural workers filled the demand. Low-cost seasonal labor and especially daily workers represented by *mardikor* groups have become common. Females recognize that seasonal work offers good short-term earnings but annualized may only equal the minimum official salary in Tajikistan. For many rural females, this income is the only money they can allocate and spend independently—usually for expenses of their children and their own personal needs including medicine. Yet such jobs do not provide security or meet the basic needs of females around hygiene, health, or childcare.

Many rural females in the study want themselves and their children to have vocational education. They consider a sewing factory, bakery, food processing facility, or factory jobs that have stable salaries and social benefits as the best alternatives to agricultural jobs.

Recommendations

(i) **Capacity development.** Develop capacities of *mardikor* groups to include operations focused on agricultural services and/or processing and exporting businesses to increase and solidify the gains of female wage workers in the value chain. The general acceptance of *mardikor* groups indicates their stability and established institutional strength although they may require additional organization and resources.

(ii)   **Promote natural farming.** Green agriculture can be considered a means for improving the situation of women and the value chain, in view of the demand for organic and natural products. Building capacity to raise awareness, knowledge, and entrepreneurial skills of women in this specific area could help reframe the traditional role of rural females as household caretaker to one of caretaker of business and the environment—becoming an "environmental manager."

(iii)   **Encourage diversification of activities.** Promote alternative agricultural activities that could offer better income opportunities for *mardikor* groups such as specialized production, harvest and postharvest phase work on crops or fruit trees, including pruning, picking, sorting, grading, and packing.

(iv)   **Provide infrastructure.** Establish greenhouses for growing high-value cash crops beyond the growing season, which could become profitable for rural households and provide formally registered paid jobs with social security.

(v)   **Support in child care.** Establish private or public daycare or kindergartens for females involved in agricultural work, especially for those who work outside their villages. These can be organized as a social entrepreneurship initiative and can be based on membership fees and support of the government or farmers' unions.

Appendix 1

Description of Study Stages II and III

Stage II

To obtain meaningful evidence on the tasks for Stage II, data collection was accomplished through two methods: a quantitative study (household survey) in irrigation systems and districts identified by the Asian Development Bank (ADB) and the Agency for Land Reclamation and Irrigation (ALRI), and qualitative method instruments such as focus group discussions and individual, key informant interviews. In total, 70 interviews and focus group discussions were held and 15 water users associations (WUAs) were studied (Table A1.1). Selection of interviewees for qualitative interviews and focus group discussions started with the identification of WUAs that are "purely" under irrigation and drainage (I&D) systems' services and various groups involved in water provision, agricultural units, and their workers.

Table A1.1: Overview of Types of Qualitative Methods, Respondents, and Locations

Semi-Structured In-depth Interviews	Yovon	Khuroson	Abdurahmoni Jomi	Jayhun	Total Interviews	Female	Male
ALRI district branch	1	1	1	5	8		8
WUA staff	4	3	2	6	15	1	14
Dehkan farmers	4	10	0	5	19	11	8
Wage workers	3	1	0	3	7	7	
Kitchen gardens	2	1	0	2	5	5	
Total	**14**	**16**	**3**	**21**	**54**	**24**	**30**
Focus groups							
WUA staff	1	1	0	1	3		3
Dehkan farmers	1	2	0	1	4		4
Wage workers	1	1	0	1	3	3	
Total	**3**	**4**	**0**	**3**	**10**	**3**	**7**

continued on next page

Table A1.1 *continued*

Expert Interviews			Total	Female	Male
ALRI	Bakhrom Gafurzoda	Deputy director	1	0	1
ALRI	Dilshod Kimsanov	Senior specialist	1	0	1
ALRI	Shahlo Sodatsarova	Women committee member, Economist	1	1	0
ALRI	Nagis	Women committee member, hydrologist	1	1	0
ALRI	Husnoro Saidova	Pump station service specialist	0	1	0
ADB Pyanj project	Anvar Murodov	Manager	1	0	1
FAO, Tajikistan	Dalerjon Domullodzhanov	National technical officer, land and water	1	0	1
		Total	**6**	**3**	**4**

ADB = Asian Development Bank, ALRI = Agency for Land Reclamation and Irrigation, FAO = Food and Agriculture Organization of the United Nations, WUA = water users association.
Source: Asian Development Bank.

All qualitative interview respondents who did not want to be named were anonymized and assigned codes to represent their identity as farmers or workers, male, female, etc.

Stage III

Data collection methods included all WUAs within selected I&D systems except WUA Kolkhozobod in the Rumi district:

(i)   **Research coverage.** In the Right Tributary I&D system: three WUAs in Yovon, four in Khuroson, and two in A. Jomi districts were covered. In the Kumsangir I&D system, seven WUAs in Jayhun districts were covered in the study.

(ii)   **Sample size.** The sample consisted of 323 households

(iii) **Sample design.** The purposive, non-probability two-stage sampling was used to design the sampling plan. In this design, the number of clusters or WUAs was determined at the first stage (Table A1.2), and households within each of the WUAs were selected randomly by the "route method" at the second stage.

Table A1.2: Sampling Based on Survey Results by Gender

WUAs	Female	Male	Total
Kumsangir I&D system	**43**	**90**	**133**
Bahoriston-2013	7	12	19
Buston-K	4	15	19
Ehyo-2013	4	15	19
Huseyn Ayub	4	15	19
Obi Ravon	13	6	19
Obrason-K	5	14	19
Qumsangir-2013	6	13	19
Right Tributary I&D system	**67**	**123**	**190**
Chorgul-2012	4	15	19
Elok Norin	6	13	19
Gulobod-2013	11	8	19
Istiklol-2010	9	10	19
Istiklol PL6	9	10	19
Mekhnat-2013	2	17	19
Navobod-2016	10	9	19
Norin	4	15	19
Obi Yovon	8	11	19
Shabnam	4	15	19
Total	**110**	**213**	**323**

I&D = irrigation and drainage, WUA = water users association.
Source: Asian Development Bank.

Since the study required sex-disaggregated data, it was decided to follow the national statistics on male–female proportion of the present population in Khatlon province. Since the study is also interested in the question of productivity, the national statistics were followed, where, in Khatlon province, there are approximately 17.5% of registered female *dehkan* farmers. Rural households with female de facto heads of household, due to outmigration of male members, abandonment, or other reasons, might not consider themselves decision makers due to stigma and/or patriarchal cultural setting. Therefore, the study did not only follow the rule of women calling themselves heads of household, as it would not be precise if only widowed women who call themselves female-headed households were chosen. Therefore, the sampling followed the following logic:

(a)   17 WUA selected in each I&D scheme, and head, middle, and end areas were considered;

(b)   54% of all respondents should be female, based on the proportion of female population in Khatlon province;

(c)   40% of all respondents should be *dehkan* farms, of which 20% should be female,[1] based on the proportion of *dehkan* farms (separately for female) in total number of farming entities in Khatlon province; and

(d)   the random walk method will be used to randomly choose the households.

The questionnaire was programmed in the application for surveys CSPro 7.1 and designed for Android OS-based tablets.

(iv)   **Choice of household units.** The choice of survey locations within each WUA village was divided into three: beginning, center, and end in relation to the water source (irrigation canal), used as survey points. A total of 19 interviews in each WUA was distributed between: head (6), middle (6), and tail end (7).

Households for the survey were selected using the random route method. The method comprises the route given to the interviewers that they should follow. Usually, in rural areas where there are no streets, the reference point for surveys will be the village administrative center (*jamoat*), shop, school, health care point or mosque, which is located in the center of a village. After finding one of these buildings, an interviewer should move along its right side.

Selection of households was conducted through a systematic door-to-door approach. For this survey, the selection step was three households. A household member who is most informed about family, its decision-making, and budget management was selected for an interview. Respondents were also chosen based on their use of agricultural land and involvement in crop production.[2]

If no adults were present in the household during the first visit, interviewers moved to the next house. If the interviewer was not able to find and conduct the interview with the initially selected households up to three times (different times of day), an interviewer moved to the next eligible household in the route.

(v)   **Survey preparation and enumeration.** There were 12 enumerators involved in conducting interviews, including two supervisors and a database person. The enumerators and field supervisors were trained and piloting of the questionnaires was conducted during 11–13 December 2019.

In all, 13 persons came from the relevant districts of the Khatlon province to participate in the trainings. The supervisors and enumerators were preselected for the study and were trained on the correct completion of the questionnaire to ensure data quality. Enumerators and supervisors also learned how to conduct interviews, how to approach the respondents, and how to hold a neutral position during the interviews.[3]

---

[1]   Note that the questionnaire does not exclude or lose the information of households who are also *dehkan* farms, meaning that the survey will also be getting their household data.
[2]   Criteria on having (use rights) irrigated lands and using them for crop production were included in the beginning of the survey questionnaire.
[3]   Baseline survey enumerators' training report is available in addition to this report.

There were some issues around finding new respondents due to refusals and replacement of enumerators following mistakes in the selection process and errors in filling out the questionnaires (Table A1.3). These issues influenced the timing and deadlines of the survey and the study.

Table A1.3: Table of Respondent Replacements

District	Sample Size	Replacement	Reasons of Households Replacement		
			Refusal	Do Not Use the Land	Did Not Fit the Quota
Khuroson	76	24	19	2	3
Jayhun (Kumsangir)	133	42	23	12	7
A. Jomi	38	24	15	5	4
Yovon	76	21	17	2	2
TOTAL	**323**	**111**	**74**	**21**	**16**

Source: Asian Development Bank.

(vi)   **Collection of questionnaires.** All finished questionnaires were submitted at the end of each day to the main FTP server. Daily collection was necessary for ongoing monitoring of the quality of completed questionnaires. The database was exported using the "Export Data" function in the CSPro 7.1 software. When exporting data, the SPSS format was selected. After processing and cleaning the data, conversions were made into Excel and Stata formats. All answers of respondents were checked for logic, value, reality, and missing data. The first stage of database verification started during data collection. Every day after receiving the questionnaires, the data were checked by a team supervisor. The bulk of the data verification was carried out after fieldwork completion.

Global positioning system (GPS) location coordinates of the households were collected during the interviews and the maps were produced with QGIS version 2.18.9 software based on Google Maps. The red dots stand for individual respondents in each I&D system (Figure A1).

Figure A1: Interview Locations for Right Tributary and Kumsangir Irrigation and Drainage Systems

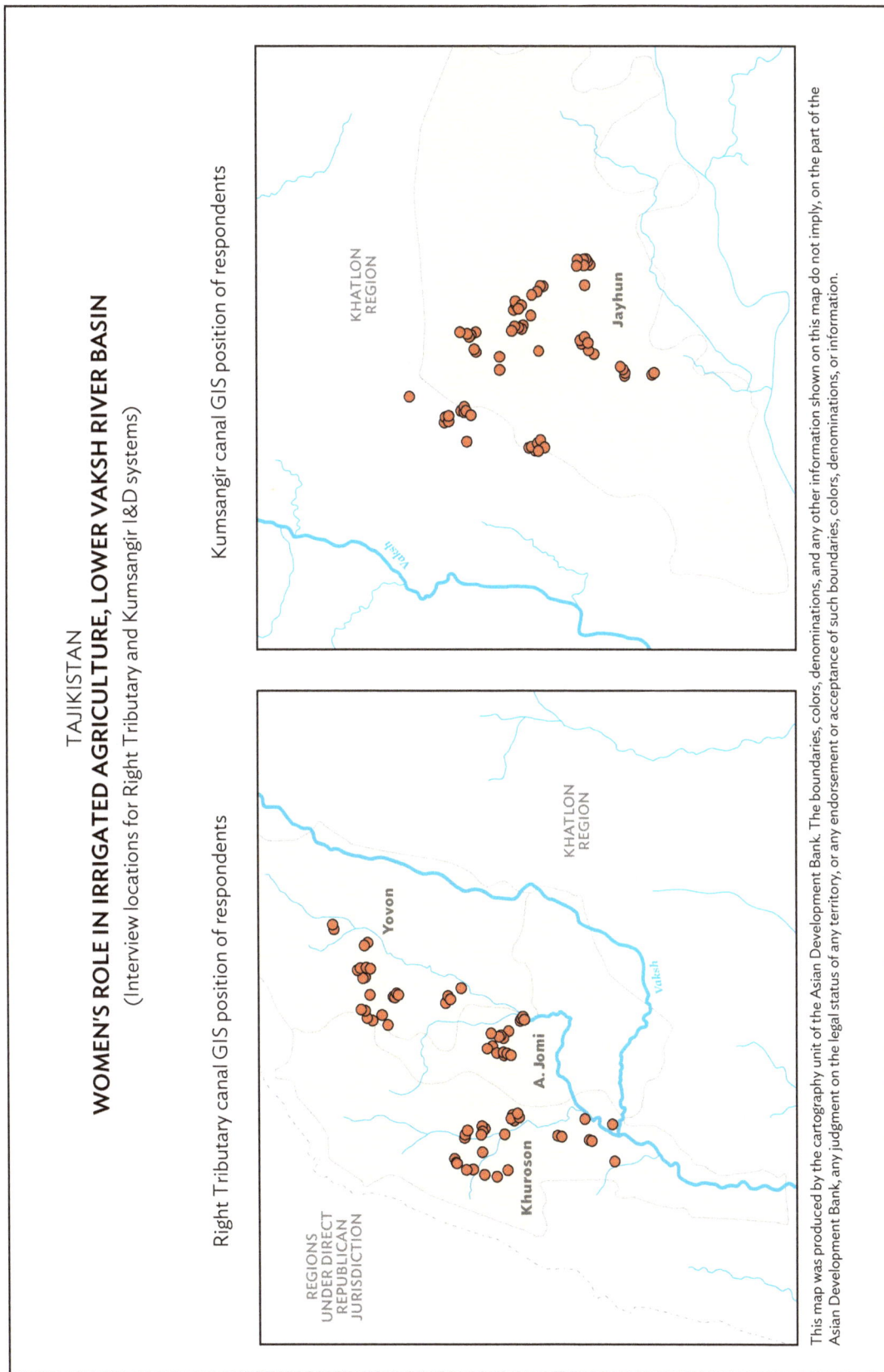

TAJIKISTAN

WOMEN'S ROLE IN IRRIGATED AGRICULTURE, LOWER VAKSH RIVER BASIN

(Interview locations for Right Tributary and Kumsangir I&D systems)

Right Tributary canal GIS position of respondents

Kumsangir canal GIS position of respondents

REGIONS
UNDER DIRECT
REPUBLICAN
JURISDICTION

Yovon

A. Jomi

Khuroson

KHATLON
REGION

Vaksh

KHATLON
REGION

Jayhun

Vaksh

200908 ABV

This map was produced by the cartography unit of the Asian Development Bank. The boundaries, colors, denominations, and any other information shown on this map do not imply, on the part of the Asian Development Bank, any judgment on the legal status of any territory, or any endorsement or acceptance of such boundaries, colors, denominations, or information.

Appendix 2

Data on Hydrology of Agency for Land Reclamation and Irrigation for Khatlon Province

Consolidated reports on water source intake and water use are based partially on the monitoring report of ALRI branches on administrative district levels, and differences in water sources are not accounted.

Reporting tables for the Yovon Irrigation system by Yovon, Khuroson, A. Jomi, and Jayhun districts include water intake (Table A2.1) from the water sources (not specified); provided from state water infrastructure, irrigation coverage including those areas irrigated by pump stations. Water intake is based on the demand for water by each district identified through planned intake amounts. The actual intake depends on the monthly availability of water in the sources, but also on discharge capacity of state water infrastructure.

Table A2.1: Water Intake in the Study Districts

	Regions and Towns	Total Water Intake from the Water Sources ('000 m³)			Total Water Provided from State Water Infrastructure ('000 m³)		
		Plan	Actual	Portion (%)	Plan	Actual	Portion (%)
1	Jayhun (September)	30,338	48,027	158	20,538	34,251	167
	Accumulated annual total	250,054	334,228	133	178,769	221,467	124
2	Yovon (September)	29,886	31,665	106	20,166	20,138	100
	Accumulated annual total	256,226	244,423	95	162,446	149,936	92
3	Khuroson (September)	10,434	14,823	142	8,244	10,494	127
	Accumulated annual total	134,065	109,792	82	104,202	81,746	78
4	A. Jomi (September)	17,419	40,179	231	6,198	19,410	313
	Accumulated annual total	203,232	303,168	149	140,798	213,248	151
5	Bakhoriston, main management office	4,695	4,808	102	4,079	3,652	90
	Total	**43,139**	**40,929**	**95**	**32,033**	**31,210**	**97**

m³ = cubic meter.

Source: Agency for Land Reclamation and Irrigation (ALRI). 2019. Report of Inventory of General and Project Focus Water Users Associations. Compiled by the author.

According to ALRI data, not all of the irrigated lands (Table A2.2) registered in the land registry (cadaster) of Land Committee are actually irrigated. Both gravity and machine (pump) irrigated areas are under the management of ALRI branch water service providers.

Table A2.2: Irrigated Area (Gravity and Machine) in the Study Districts

	Regions and Towns	Irrigated Land Area[a] (ha)	Portion Coverage of Irrigated Lands (%)	Irrigated Land Area Covered by ALRI (ha)	Including Machine-Irrigated (ha)	Portion of Machine-Irrigated from Total Irrigated Area Covered by ALRI (%)
1	Jayhun	28,197	67	18,773	8,498	45
2	Yovon	26,824	92	24,624	14,147	57
3	A. Jomi	19,048	81	15,473	4,579	30
4	Khuroson	10,632	88	9,388	2,681	29
5	Bakhoriston, main management office	0		4,380		

ALRI = Agency for Land Reclamation and Irrigation, ha = hectare.
[a] According to the data of the Land Committee for 1 January 2019.

Source: Agency for Land Reclamation and Irrigation (ALRI). 2019. Report of Inventory of General and Project Focus Water Users Associations. Compiled by the author.

ALRI branches consolidate the data provided by WUAs (Table A2.3) based on demand and availability (discharged amounts of water). The price for gravity water provision is calculated based on irrigated territories registered as service areas of ALRI branches including WUAs. Pumped water price service is calculated based on electricity use by the pump stations and their maintenance.

Table A2.3: Water and Electricity Amounts and Total Prices in the Study Districts

	Regions and Towns	Water Provided for Irrigation (million m³)	Price of Water Provision (TJS'000)	Electricity Use ('000 kW)	Electricity Price (TJS'000)
1	Jayhun	177.0	3,539.0	28,793.7	2,074.6
2	Yovon	166.2	2,939.8	52,385.4	3,651.1
3	A. Jomi	168.4	1,820.0	514.7	35.7
4	Khuroson	68.5	1,227.2	19,287.5	942.0
5	Bakhoriston, main management office	27.1	514.5	0	0

kW = kilowatt, m³ = million cubic meter, TJS = somoni.

Source: Agency for Land Reclamation and Irrigation (ALRI). 2019. Report on the Monitoring of Water Users Association (WUAs). Compiled by the author.

Characteristics of Selected Canals

Canal	Right Tributary (Pravaya Vetka)	Kumsangir
Location	**Khatlon province:** Yovon, Khuroson, Abdurahmoni Jomi districts	**Khatlon province:** Jayhun, Rumi districts
Command area	32,495 ha	7,775 ha
Constructed in	1961 to 18 May 1968	1932 to 1933 (1968 in Rumi)
Operator	Yovon, Khuroson, Abdurahmoni Jomi ALRI branches (Vodhoz)	Jayhun, Rumi ALRI branches (Vodhoz)
Discharge capacity: Maximum	50 m³/s	50 m³/s
Normal	5–28 m³/s (5m³/s in the tail end in the A. Jomi ALRI branch)	32–38 m³/s
Length	50 km	50 km
Regulating facilities	110 structures (in the project area)	23 structures
Land area used for canal construction (only project area)	130.7 ha	367.8 ha
Present balance cost of the canal	TJS965,385.7	TJS1,316.2
Total irrigation network	**749.0 km**	**582.4 km**

ALRI = Agency for Land Reclamation and Irrigation, ha = hectare, km = kilometer, m³/s = cubic meter per second, TJS = somoni.

Source: Agency for Land Reclamation and Irrigation (ALRI). 2019. Report on the Monitoring of Water Users Association (WUAs). Compiled by the author.

Appendix 4

General Demographics and Characteristics of Project-Relevant Water Users Associations

Characteristics of Members	Total for Both Schemes	Right Tributary Canal	Kumsangir Canal
Total number of people	203,845	121,523	82,322
Total number of households	28,497	16,799	11,698
Total number of female-headed households	1,925	1,152	773
Number of WUA members	4,821	2,338	2,483
Number of WUA members, female	418	291	127
Total number of *dehkan* farms in WUAs	5,094	2,611	2,483
Dehkan farms led by women	425	298	127
Number of *dehkan* farms, <30 ha	5,054	2,582	2,472
Number of *dehkan* farms, >30 ha	40	29	11
Total irrigated area (ha)	32,989	20,209	12,780
Dehkan farms (ha)	26,340	16,821	9,519
Presidential lands (ha)	1,809	934	875
Household plots (ha)	4,656	2,270	2,386
Other (ha)	184	184	0
Area of irrigated land *dehkan* households	26,340	16,821	9,519
Area of *dehkan* farms and households, <30 ha	22,376	13,617	8,759
Area of *dehkan* farms and households, >30 ha	3,964	3,204	760

ha = hectare, WUA = water users association.

Source: Author's analysis and compilations based on monitoring reports of WUAs of Yovon, A. Jomi, Khuroson, and Jayhun districts.

Appendix 5

Mardikor Groups—The Story of Apai Guli

Apai Guli is a *mardikor* group leader in the Kumsangir I&D system.[4] She is over 50 years old. She has two children and her eldest son left their home to work in the Russian Federation.

I [have been] a leader of [a] *mardikor* group since 2003. I was 30 years old at that time. Before that, I worked as a *mardikor* leader on a farm, then the farmlands were taken away and now I work as a *mardikor*. I do not have any other job, nor officially registered as being employed. During those years, we used to travel to Vijdan. At that time, such jobs were not offered in our territory. Earlier, we worked only in other villages, but now *mardikor* groups have multiplied and there is no need to go far, therefore, we work only in our village. We have already two *mardikor* groups in our *mahalla* and when there is lack of people we cooperate and share, take additional people from each other.

The work of a *mardikor* starts at 8:00 a.m. and continues [until] 6:00 to 7:00 p.m. The end of February is the time when cucurbit seeds are planted and that is the start of our work season and it ends in December. The period is 11 months but if we calculate the time purely used for work it could be around 10 months. The availability of work partly depends on the weather. If it is not cold, we can work every day. The other condition is the availability of demand and how quickly we can find customers without any interruptions.

Although women go for these activities out of necessity and it is a hard job, they earn well, they cover many household expenses and their household members, including husbands, are satisfied. The group is dynamic as many girls get married and leave to go to other villages or are occupied with their household chores. Yet, there is always a bigger pool of women who would like to take the job. When a girl or woman asks to [be accepted] to the groups, I first explain then take her to the field with the group. When she leaves her house, all the responsibility is on me and after they finish their work and go home I check whether they reached their houses. Very often, I have to organize my group in the evening, so I call all the women from my list until I get the number of people that I need. In my experience, there were no cases when men (husbands) came and complained to me. There were men who came and asked [me] to accept their wives and daughters to my group.

Usually, the quantity of people in a group depends on the types of activity. [It ranges] between 15 to 25 persons per group. For weeding and cleaning, one furrow (100 meters long) will cost TJS7 to TJS8. For harvesting, we take around TJS10 per kilogram and the lunch is always offered by the farmer who hired us. We are usually offered work for weeding, tilling the land for carrots, onions, and melons. Cotton-related activities have their own workloads and costs. I have my own registry book; I also keep the old ones.

---

4   Pseudonyms are used to protect the identities of the respondents in the study area.

I negotiate with farmers first also on the price if we both agree then we start the work. I also try to check the character of the farmer (male) if he behaves rude or unrespectful, then I ignore the offer from him. In general, most of the farmers with whom we work [...] become our constant clients. This also means that they do not give out this work to other *mardikor* groups.

I usually first do the work and then take the fee at the end of each day. The transportation is organized by the farmers.

My husband and I did not have shares, and we have been sub-renting land from other farmers for around TJS5,000 to TJS7,000 per hectare. People usually do not rent out good land and you never know what you will get at the end even if you do everything correctly and buy good seeds. We have not been lucky with planting melons on our land [so] we changed and sub-rented other lands and decided to plant carrots. We harvested them now [and] offered to sell for TJS0.50 per 50-kilogram sack.

I spend the money I earn for [...] food items. I have never thought to buy a dress or something solid like gold things. We cannot afford to spend on such things. Last year, I had to use the remittance from my son to cover the debt on land rent. In case you rent, you take all the risk on your own and you have to pay the rent even if you received bad land or seeds. The money earned is spent almost immediately and I cannot save it.

Mardikor groups [is also a way for us to socialize]. While working together, women share their life stories, their daily problems. Our group organizes a get-together to celebrate *Navruz* or the New Year.

Glossary

Amelioration	Term used in the context of improving the land, especially the drainage condition of soils; sometimes also used in a broader sense to encompass both irrigation and drainage similar to the use of the term "reclamation"
Dehkan farm	Family-based small-scale enterprise that produces and markets agricultural products using labor of family members on a land plot transferred to the head of family for lifelong inheritable ownership, registered or not registered officially as a legal entity (*dehkan* literally means "farmer")
Jamoat	Village level state administration unit
Kitchen garden	A small plot of land adjacent to a house and used for agriculture or gardening
Kolkhozchi	A term carried from the past and translated as kolkhoz worker. Presently, this term refers to hired workers mainly involved in cotton production
Mahalla	A traditional community structure, a local neighborhood council
Mirob	A person responsible for monitoring and water delivery in irrigation systems
Mardikor or *yordamchi*	Hired daily workers
Sahmdor	Shareholder and registered employee of a *dehkan* farm

References

Asian Development Bank. 2020. *Women's Time Use in Rural Tajikistan*. Manila. DOI: http://dx.doi.org/10.22627/TCS 200167-2.

Agency for Land Reclamation and Irrigation (ALRI). www.alri.tj (accessed January 2020).

A.M. Danzer, B. Dietz, and K. Gatskova. 2013. *Tajikistan Household Panel Survey: Migration, Remittances and the Labor Market*. Regensburg.

C. Francisco and M. Bakanova. 2013. Tajikistan: Reinvigorating Growth in the Khatlon Oblast. *Europe and Central Asia Knowledge Brief* No. 70. Washington, DC: World Bank.

Government of Tajikistan. 1994. Constitution of the Republic of Tajikistan (amended 2003).

Government of Tajikistan. 2001. Resolution of the Government of Republic of Tajikistan No. 391, dated 6.08.2001 Government programme "On main directions of government policy on providing equal rights and opportunities for man and women in Republic of Tajikistan for 2001–2010."

Government of Tajikistan. 2005. Law of Republic of Tajikistan "On state's guarantees of equality of men and women and equal opportunities for fulfillment of their rights. https://cis-legislation.com/document.fwx?rgn=8064 (accessed 17 November 2020).

Government of Tajikistan. 2006. Resolution of the Government of Republic of Tajikistan No. 496 on Co-mentoring, Selection and Job placement of Leading Personnel of the Republic of Tajikistan from Among Gifted Women and Girls for Years 2007–2016. 11 January.

Government of Tajikistan. 2020. Law on Water Users' Associations No. 1668. 2 January.

Government of Tajikistan, Committee for Women and Family. https://comwom.tj/ru/history (accessed January 2020).

Government of Tajikistan, Statistical Agency under the President. 2018. Statistical Collection on Agriculture in the Republic of Tajikistan.

International Labour Organization (ILO). 2010. *Migration and Development in Tajikistan: Outmigration, Return and Diaspora*. Moscow. September. https://www.ilo.org/moscow/information-resources/publications/WCMS_308939/lang--en/index.htm.

ILO. 2018. *Women and Men in the Informal Economy: A Statistical Picture*. https://www.ilo.org/global/publications/books/WCMS_626831/lang--en/index.htm.

E. Katz. 1995. Gender and Trade within the Household: Observations from Rural Guatemala. *World Development*. 23 (2). pp. 327–342.

R. Meinzen-Dick and M. Zwarteveen. 1998. Gendered Participation in Water Management: Issues and Illustrations from Water Users Associations in South Asia. *Agriculture and Human Values*. 15. pp. 337–345.

N. Mukhamedova and K. Wegerich. 2018. The Feminization of Agriculture in Post-Soviet [Union] Tajikistan. *Journal of Rural Studies*. 57. January. pp. 128–139.

C. Oriol. 2018. *Country Study Tajikistan*. World Bank and the Eurasian Center for Food Security.

S. Robinson et al. 2008. Land Reforms in Tajikistan: Consequences for Tenure Security, Agricultural Productivity and Land Management Practices. In R. Behnke, ed. *The Socio-Economic Causes and Consequences of Desertification in Central Asia*. Springer Science + Business Media B.V. pp. 171–203.

J. Sehring. 2006. *The Politics of Irrigation Reform in Tajikistan*. Discussion Paper No. 29, Justus-Liebig-Universität Gießen, Zentrum für Internationale Entwicklungs-und Umweltforschung (ZEU), Giessen. https://www.econstor.eu/bitstream/10419/21925/1/DiscPap29.pdf.

Statistical Agency under President of the Republic of Tajikistan. Labour Market in the Republic of Tajikistan 2017. http://oldstat.ww.tj/en/img/7b6f49435ed5ae6ec685562d6e2858 3a_1426679088.pdf (accessed January 2020).

G. Standing. 1999. Global Feminization through Flexible Labor: A Theme Revisited. *World Development*. 27 (3). pp. 583–602.

United Nations Economic Commission for Europe. 2012. *Roadmap of the National Policy Dialogue on IWRM in the Republic Tajikistan*. Dushanbe, Tajikistan.

United States Agency of International Development (USAID). 2019. Feed the Future Project. https://www.usaid.gov/tajikistan/agriculture-and-food-security (accessed August 2020).

USAID. 2010. Gender Assessment USAID/Central Asian Republics. Washington, D.C.

A.S. Wharton. 2013. *The Sociology of Gender: An Introduction to Theory and Research*. Hoboken, N. J.: Wiley-Blackwell.

World Bank. Database. Agriculture, Forestry, and Fishing, Value Added (% of GDP). https://data.worldbank.org/indicator/NV.AGR.TOTL.ZS?locations=TJ (accessed July 2019).

World Economic Forum. 2018. *The Global Gender Gap Report*. https://reports.weforum.org/global-gender-gap-report-2018/ (accessed August 2020).